MA
W

Work Out

Statistics

A-Level

The titles in this series

For GCSE examinations

Accounting
Biology
Business Studies
Chemistry
Computer Studies
Economics
English
French
Geography
German
Graphic Communication
Human Biology
Maths
Modern World History
Numeracy
Physics
Social and Economic History
Sociology
Spanish
Statistics

For A-level examinations

Accounting
Applied Maths
Biology
Business Studies
Chemistry
Economics
English Literature
French
Physics
Pure Maths
Statistics

Macmillan College Work Outs for degree and professional students

Dynamics
Electric Circuits
Electromagnetic Fields
Electronics
Elements of Banking
Engineering Materials
Engineering Thermodynamics
Fluid Mechanics
Mathematics for Economists
Molecular Genetics
Operational Research
Organic Chemistry
Physical Chemistry
Structural Mechanics
Waves and Optics

MACMILLAN WORK OUT SERIES

Work Out Statistics A-Level

A. D. Ball

and

G. D. Buckwell

First published 1986
Reprinted (with corrections) 1987, 1989, 1990
Second edition 1991

Published by
MACMILLAN EDUCATION LTD
Houndmills, Basingstoke, Hampshire RG21 2XS
and London
Companies and representatives
throughout the world

Printed in Hong Kong

British Library Cataloguing in Publication Data
Ball, A. D.
Work out statistics A-level.–(Work out series)
1. Mathematical statistics–Examinations,
questions, etc.
I. Title II. Buckwell, G. D.
519.5′076 QA276.2
ISBN 0-333-56333-6

Contents

Acknowledgements

The University of London Entrance and School Examinations Council accepts no responsibility whatsoever for the accuracy or method in the answers given in this book to actual questions set by the London Board.

Acknowledgement is made to the Southern Universities' Joint Board for School Examinations for permission to use questions taken from their past papers but the Board is in no way responsible for answers that may be provided and they are solely the responsibility of the authors.

The Associated Examining Board and the University of Oxford Delegacy of Local Examinations wish to point out that worked examples included in the text are entirely the responsibility of the authors and have neither been provided nor approved by the Board.

The authors and publishers wish to thank the following who have kindly given permission for the use of copyright material:

The Associated Examining Board, Southern Universities' Joint Board, University of Cambridge Local Examinations Syndicate, University of London School Examinations Board, University of Oxford Delegacy of Local Examinations and the Welsh Joint Education Committee for questions from past examination papers.

Longman Group for tables from *Statistical Tables for Biological, Agricultural and Medical Research* by Fisher and Yates.

Murdoch and Barnes for table from *Statistical Tables for Science, Engineering, Management and Business Studies*, Macmillan.

Every effort has been made to trace all the copyright holders but if any have been inadvertently overlooked the publishers will be pleased to make the necessary arrangement at the first opportunity.

Examination Boards for A-Level

Syllabuses and past examination papers can be obtained from:

The Associated Examining Board (AEB)
Stag Hill House
Guildford
Surrey GU2 5XJ

University of Cambridge Local Examinations Syndicate (UCLES)
Syndicate Buildings
Hills Road
Cambridge CB1 2EU

Joint Matriculation Board (JMB)
78 Park Road
Altrincham
Cheshire WA14 5QQ

University of London School Examinations Board (L)
University of London Publications Office
52 Gordon Square
London WC1E 6EE

University of Oxford (OLE)
Delegacy of Local Examinations
Ewert Place
Summertown
Oxford OX2 7BZ

Oxford and Cambridge Schools Examination Board (O&C)
10 Trumpington Street
Cambridge CB2 1QB

Scottish Examination Board (SEB)
Robert Gibson & Sons (Glasgow) Ltd
17 Fitzroy Place
Glasgow G3 7SF

Southern Universities' Joint Board (SUJB)
Cotham Road
Bristol BS6 6DD

Welsh Joint Education Committee (WJEC)
245 Western Avenue
Cardiff CF5 2YX

Northern Ireland Schools Examination Council (NISEC)
Examinations Office
Beechill House
Beechill Road
Belfast BT8 4RS

Introduction

Although intended for A-Level students following a course in statistics, first year or university students who are either meeting statistics for the first time or who require a refresher course, would find that this book contains useful support material.

How to Use this Book

The topics included in *Work Out Statistics A-Level* cover the syllabuses of all examining boards at A-Level in the subject. In particular, your attention is drawn to chapter 15 which has been included to cover those areas which are examined but are often not contained in textbooks. The chapter headings follow the systematic order offered by most contemporary textbooks. Most topics contain worked solutions of past paper questions from examining boards. As different boards and papers take the subject to varying levels, so some topic areas will prove to be quite difficult. Such topics and worked examples have been marked with an asterisk (*) and you need not work through these sections if they are not given particular emphasis on your syllabus.

Revision

The course you are following will normally be a subset of the contents of this book. It is therefore important to know your syllabus. (Syllabuses and past papers can be obtained from the examination boards listed on pp. ix–x). Once you have eliminated the chapters/topics which you do not require, revise a topic at a time trying the appropriate worked examples and then comparing your answer with the worked solution. The exercises at the end of each chapter have been carefully chosen to cover fully the subject area in that chapter, so try to answer as many of these as possible before referring to the answers at the back of the book. Included in the answer section are some hints on the solution of each exercise; these should help you to understand where you are going wrong if your answer is different from the one given by the authors.

Most numerical solutions in this book have been worked using four-figure accuracy. Slight numerical differences may exist between your answers if you have worked to a higher degree of accuracy. However, the authors feel that for examination questions which are to take less than 30 minutes each, a higher degree of accuracy is time consuming and unnecessary for the purpose of solving the problem.

It is important that you should familiarise yourself with the statistical tables that you are to use during the examination as they may vary from one examination board to another.

Remember to start your revision in good time and use this book as a positive aid to the active and thorough knowledge of your statistics course.

1 Presentation of Statistics

1.1 Basic Definitions

Statistics is concerned with the collection of *data*, its organisation, presentation and analysis. In this first chapter, we will consider the organisation and presentation. To investigate the characteristics of a particular group of individuals or objects called the *population*, a *sample* is taken, and the values of the *variable* to be investigated, say the heights of people in England, are taken.

A population can be *finite* or *infinite*.

The population of this country would be finite.

The decimal numbers between 0 and 1 would be infinite.

The variable which is being measured can be *discrete* or *continuous*.

For example, the number of children in a family can only take the values {0, 1, 2, 3, . . .}. It is discrete. However, the height of a person can take any decimal value over a certain range and is therefore continuous.

1.2 Tabulation and Diagrams

Once the data has been gathered, it can be represented by a *graph*. The main types of graph in use are *bar graphs, pie graphs, pictographs* or *line graphs*.

The number of rooms in each of 40 houses in a particular street is given by the following set of data:

5 6 4 3 3 6 6 4 5 4 7 8 3 5 4 4
4 8 8 3 5 5 6 5 7 4 6 5 4 3 3 4
5 5 4 7 6 10 9 8

In order to make the information manageable, we divide it into *groups*. The information is recorded in a table called a *frequency table*. The number of times a value occurs is called its *frequency*. The recording of the values is called a *tally* (see Table 1.1).

Table 1.1

No. of rooms	*Tally*	*Frequency*
3	~~1111~~ 1	6
4	~~1111~~ ~~1111~~	10
5	~~1111~~ 1111	9
6	~~1111~~ 1	6
7	111	3
8	1111	4
9	1	1
10	1	1

The information can now be presented as in Table 1.2. Note this is a *discrete distribution*. Table 1.2 represents the *frequency distribution.*

Table 1.2

No. of rooms	3	4	5	6	7	8	9	10	Total
Frequency	6	10	9	6	3	4	1	1	40

(a) Bar Graph

Figure 1.1 shows two ways of representing the information in a bar graph. Make sure that the axes are clearly labelled.

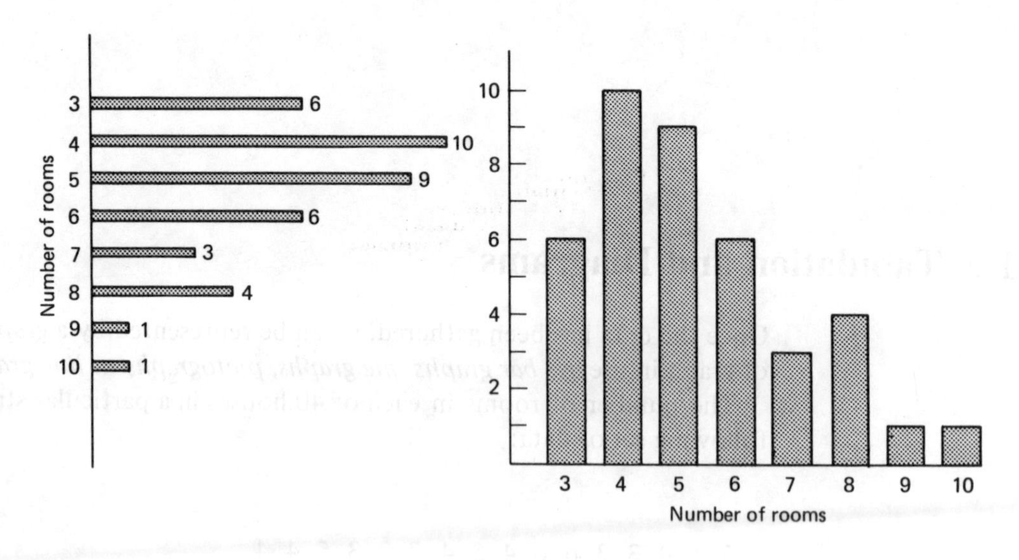

Figure 1.1

(b) Pie Graph

In a pie graph, a circle is divided up into sectors to represent each value of the variable. Since there are 360° in a circle, and this is representing 40 houses, each house will correspond to $360° \div 40 = 9°$. The angles are given in Table 1.3.

Number of rooms in a street of 40 houses

Figure 1.2

Table 1.3

No of rooms	3	4	5	6	7	8	9	10	Total
Angle	54	90	81	54	27	36	9	9	360

The pie chart can be seen in Fig. 1.2. Each sector must be clearly labelled.

If pie charts are used to compare two or more sets of values, then the radii of the circles are chosen so that the ratios of the areas of the circles are the same as the ratio of the total quantities they represent.

For example, if a second street containing 50 houses was chosen, then if the radius of the first circle is $R_1 = 4$ cm, and R_2 is the radius of the second circle, then

$$\frac{\pi R_1^2}{\pi R_2^2} = \frac{40}{50} = 0.8$$

$$\Rightarrow R_1^2 = 0.8\, R_2^2$$

But $R_1 = 4$ $\quad \Rightarrow \dfrac{4^2}{0.8} = R_2^2 \quad \Rightarrow R_2 = \sqrt{\dfrac{4^2}{0.8}} = 4.47$ cm

The radius of the larger circle should be 4.47 cm.

(c) Pictographs

A pictograph uses the concept of an eye-catching shape to represent a certain quantity. Consider the example in Fig. 1.1 again and also as shown in Fig. 1.3. This diagram has good visual appeal, but part shapes are sometimes difficult to read. In Fig. 1.3, it is obvious that 1 house is represented by half of the shape.

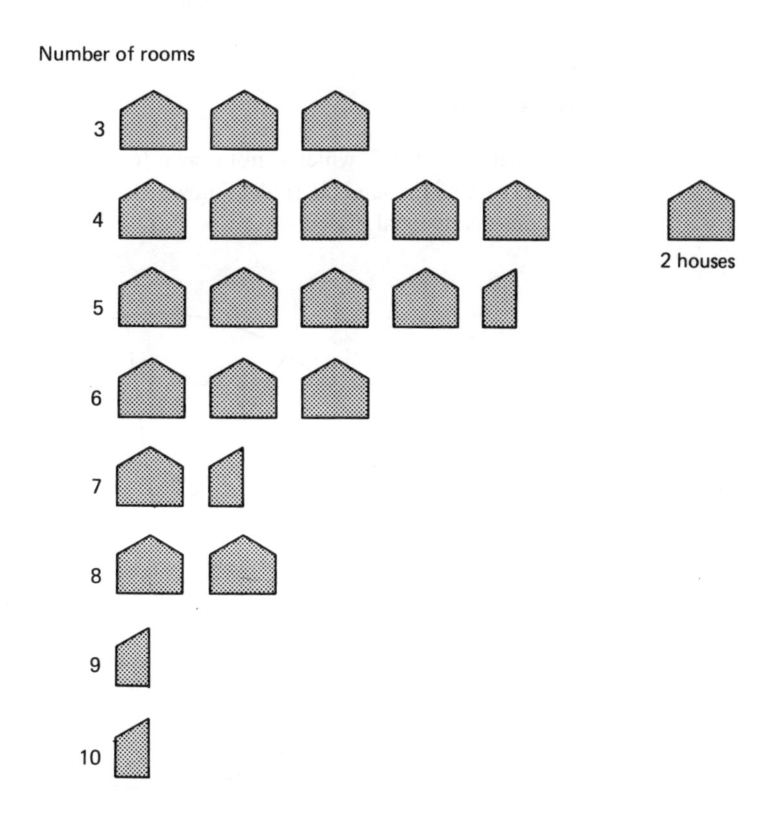

Figure 1.3

(d) Line Graph

Figure 1.4 gives once again the results of Table 1.1 but shown in a different way. This is a very simple use of this type of graph; they are usually used for *grouped frequency* situations.

A grouped frequency table is given in Table 1.4.

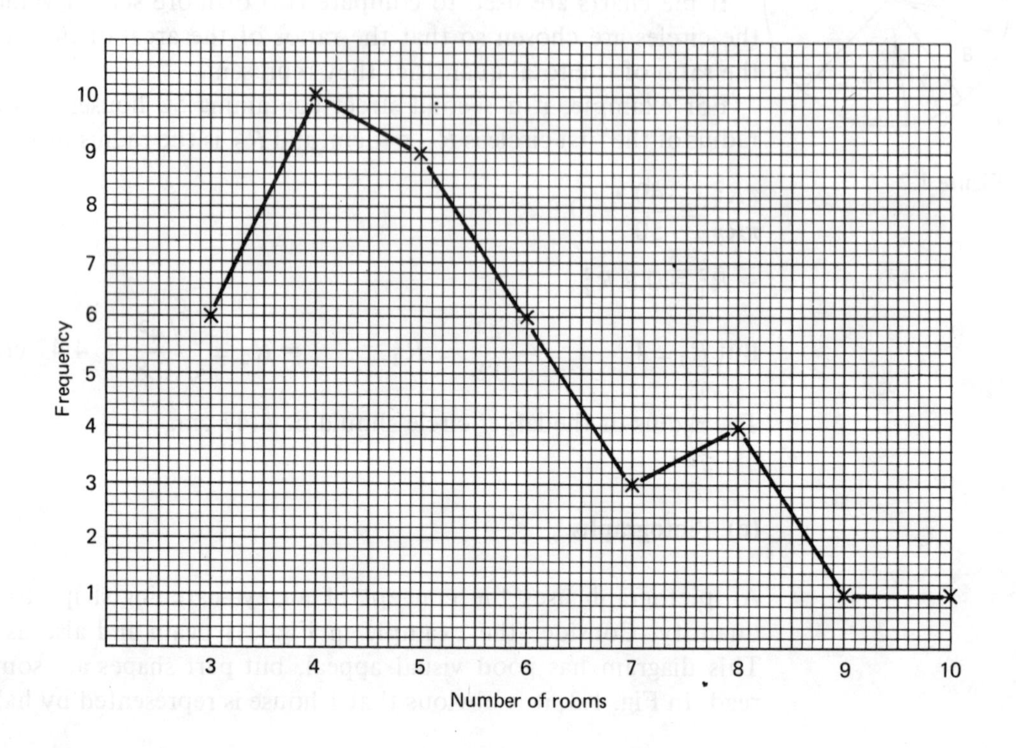

Figure 1.4

Worked Example 1.1

The pie chart in Fig. 1.5, which is not drawn to scale, shows the distribution of types of land in a small part of England. It is also known that the areas covered by marshland and forest can be considered equal. Calculate

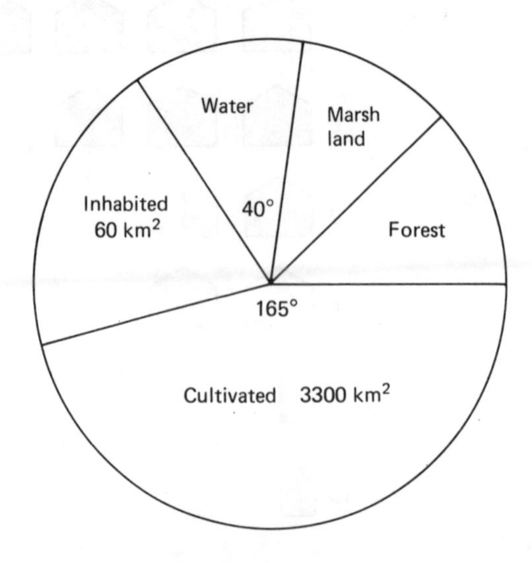

Figure 1.5

(i) the area covered by water;
(ii) the angle for the inhabited sector;
(iii) the total area of land surveyed;
(iv) the area of forest.

Solution 1.1

The important point here is that 3300 km^2 of cultivated land is represented by 165°.

$$\Rightarrow 1^\circ \text{ represents } \frac{3300}{165} = 20 \text{ km}^2$$

(i) Area covered by water = 20 x 40 = 800 km^2
(ii) Angle for inhabited sector = $\frac{60}{20} = 3^\circ$
(iii) The total area = 360 x 20 = 7200 km^2
(iv) The marsh land and forest sections have a total angle of 360° – (165° + 3° + 40°) = 152°
The forest is half, which is 76°
⇒ The area of forest = 76 x 20 = 1520 km^2

1.3 Continuous Distributions, Histograms

The lengths of 20 seedlings were measured in a growing experiment, and the results, given to 2 decimal places were as follows (all in cm).

4.61	3.84	4.79	6.01	5.43	3.84	4.62
6.20	4.85	5.34	6.71	4.21	3.80	5.14
6.04	5.18	4.27	4.63	4.38	5.49	

This information is grouped together in *classes*. Practice will decide the best classes to use. If the classes are too small or large, the resulting graphs will often be useless. The resulting diagram must show the *distribution*.

The smallest value is 3.80 cm.
The largest value is 6.71 cm.

A class width of 0.5 cm gives 6 classes.

Referring to Table 1.4, 3.75 – 4.24 is called the *class interval.* 3.75 is the *lower class limit*, and 4.24 the *upper class limit.*

Table 1.4

Class (cm)	*Tally*	*Frequency*
3.75-4.24	1111	4
4.25-4.74	~~1111~~	5
4.75-5.24	1111	4
5.25-5.74	111	3
5.75-6.24	111	3
6.25-6.75	1	1

The *class boundaries* are taken as 3.745 cm, 4.245 cm, 4.745 cm, etc. These values are used because the distribution is *continuous* and although values are given to 2 decimal places, and 3.745, etc., cannot occur, these class boundaries avoid any confusion as to where to put a particular value.

For the class interval 5.25 – 5.74, the *upper class boundary* is 5.745 and the *lower class boundary* is 5.245.

The *class width* is the difference between the upper class boundary and the lower class boundary.

Hence width = 5.745 – 5.245 = 0.5

A bar chart for a continuous distribution is usually referred to as a *histogram*. The height of the bar in a histogram does not always give the frequency of that class (see Fig. 1.7). The results are shown in Table 1.5 with unequal class widths.

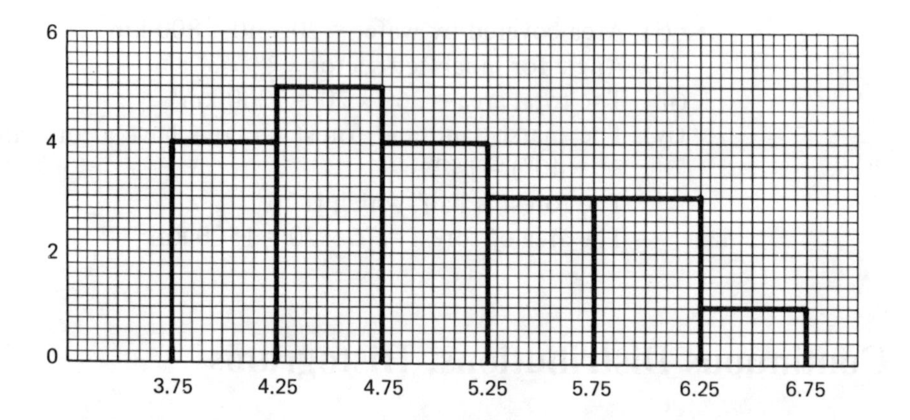

Figure 1.6

Table 1.5

Class (cm)	*Tally*	*Frequency*
3.75–4.74	~~1111~~ 1111	9
4.75–5.24	1111	4
5.25–6.24	1111	4
6.25–6.75	1	1

The histogram has the property that the areas essentially give the frequency. Hence in Fig. 1.7 the width of the first block is 2 units, its height is 4.5 hence area = 2 x 4.5 = 9. This is the true frequency for the 3.75 – 4.74 class.

In many diagrams, the vertical axis on a histogram is not labelled. It should not be labelled frequency, but can be labelled frequency density as in Fig. 1.7.

1.4 Frequency Polygons and Curves

Consider the information given in Table 1.6 for the distribution of student ages in a student population. The *mid-point* of the class is the average of the upper and lower class boundaries. For the first class it equals

$$\frac{14.95 + 15.95}{2} = 15.45$$

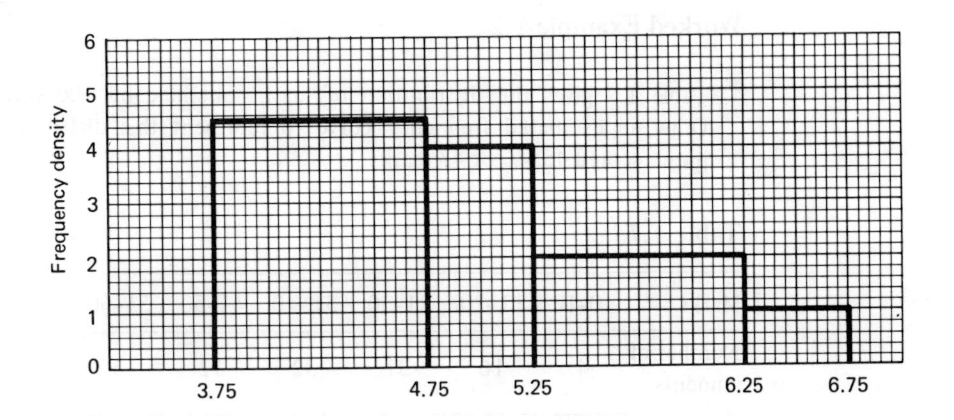

Figure 1.7

Table 1.6

Age (years)	*Class mid point*	*Number of students*
15-15.9	15.45	15
16-16.9	16.45	23
17-17.9	17.45	31
18-18.9	18.45	16
19-19.9	19.45	8
20-20.9	20.45	4

The others are 16.45, 17.45, etc. The frequencies are plotted at these points, as in Fig. 1.8. Note that the polygon is drawn down to the horizontal axis.

The histogram is also shown so as to indicate the relationship between the two. The value of a frequency polygon is that more than one set of values can be compared on the same diagram. If the widths of the classes are unequal, the height of the points is reduced in the same way as the plotting of a histogram.

If the class widths are small, then the polygon will look approximately like a curve. This is called the *frequency curve*, or *distribution curve.* An approximation to this curve can be drawn as illustrated in the following example.

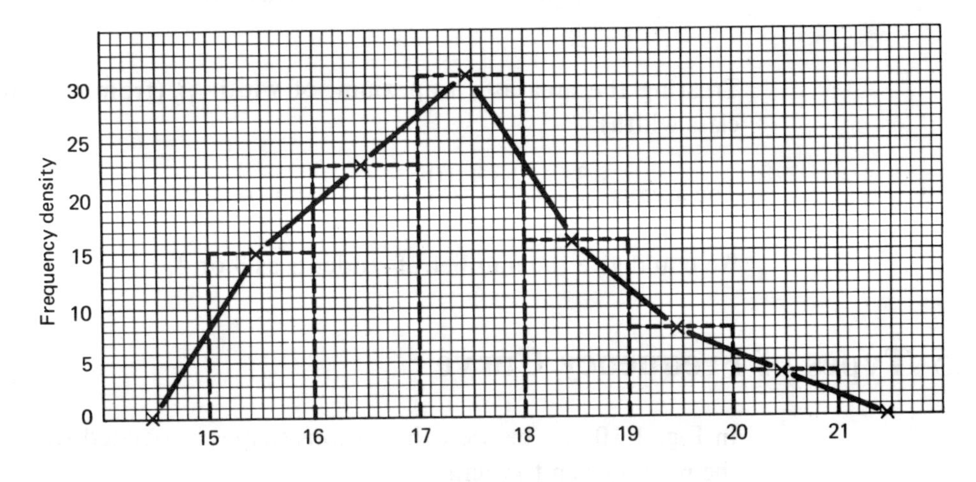

Figure 1.8

Worked Example 1.2

Table 1.7 gives the distribution of marks obtained by 100 students in an examination. Draw an estimated frequency curve to represent this distribution. Estimate how many pupils scored 45, 46 or 47 marks.

Table 1.7

Marks	0-10	11-20	21-30	31-40	41-50	51-60
Number of students	4	16	31	32	12	5

Solution 1.2

The mid-points of the classes are 5, 15.5, 25.5, etc. (note the class boundaries for the first class are −0.5 and 10.5). The distribution curve is shown in Fig. 1.9. It is important to realise at this stage that it is the area under the curve that gives the total number of students as 100, in exactly the same way as the histogram, treating the class width as 1 unit. To find the number of students who scored 45, 46 or 47, we need to find the shaded area shown in the diagram.

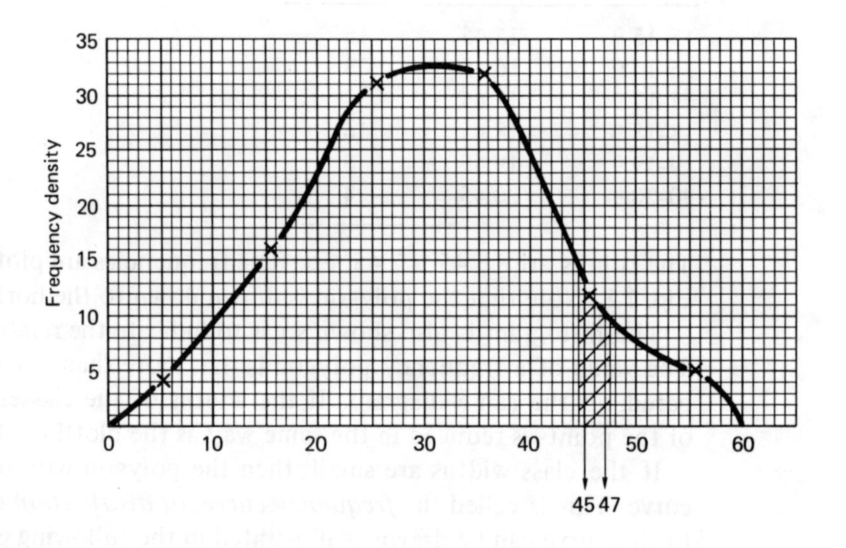

Figure 1.9

Treating the area as a trapezium, of width $\frac{3}{10}$ (from 44.5 to 47.5 is 3/10 of the class width)

$$\text{The area is } \frac{3}{10} \times \frac{13 + 10}{2} = 3.5 \text{ (approx.)}$$

We could give the number of students as 3 or 4.

1.5 Types of Frequency Curve

In Fig. 1.10 can be seen the various shapes associated with frequency curves, and the names given to them.

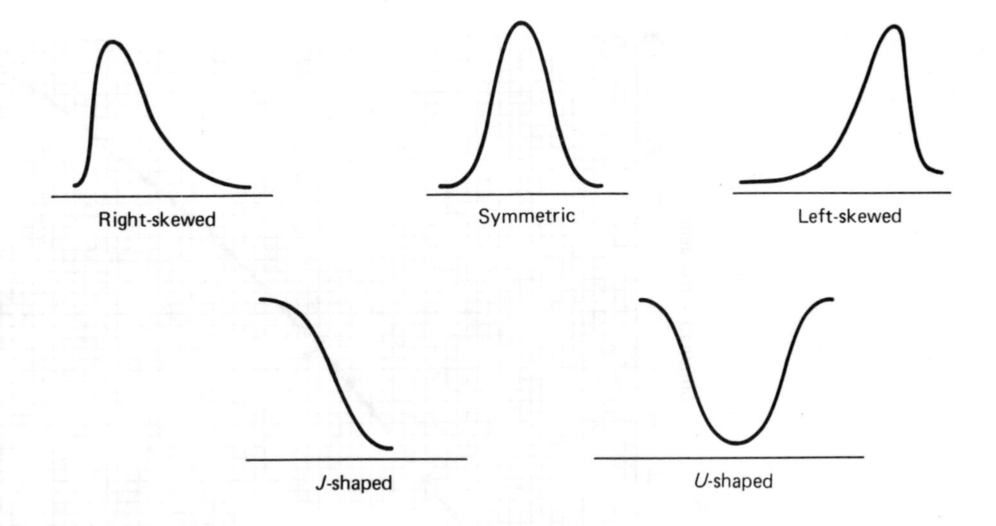

Figure 1.10

1.6 Cumulative Frequency

The total frequency up to a particular value or upper class boundary, is called the *cumulative frequency*. Consider again the values obtained in Table 1.4.

Table 1.8

Class	*Frequency*	*Cumulative frequency*
3.75-4.24	4	4
4.25-4.74	5	9
4.75-5.24	4	13
5.25-5.74	3	16
5.75-6.24	3	19
6.25-6.75	1	20

The cumulative frequencies have been added in Table 1.8. These results can now be plotted in a *cumulative frequency curve*, or *ogive*. The points should be plotted at the upper class boundaries, in this case 4.245, 4.747 etc. Accuracy prevents this, and hence they are plotted at 4.25, etc. (see Fig. 1.11).

1.7 Exercises

Exercise 1.1

State with reasons, whether the following types of data are continuous or discrete.

(i) The temperature of the sea in summer.
(ii) The fares on the London Underground.
(iii) The number of children in schools.
(iv) The lifetime of a light bulb.

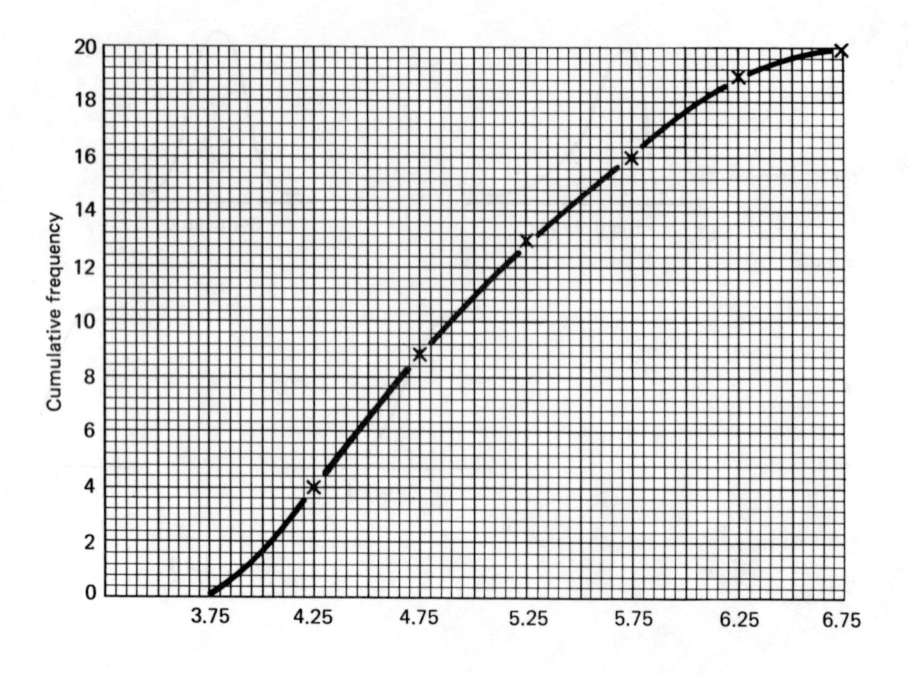

Figure 1.11

Exercise 1.2

The following data are the marks of students in an examination.

24	61	36	81	45	37	78	65	60	39
47	51	35	82	63	60	58	37	52	24
90	87	18	8	29	46	35	64	27	50

(i) Draw a tally chart using classes 0–9, 10–19, etc.
(ii) Draw a frequency distribution for the data.

Exercise 1.3

The data in Table 1.9 shows the age distribution of a small village.

Table 1.9

Age (years)	*Frequency*
0–14	18
15–19	21
20–24	38
25–34	41
35–44	38
45–59	15
60–	20

Draw a histogram to represent this information, stating any assumptions you make.

Exercise 1.4

Table 1.10 shows the distribution of cloud type in Brighton during August 1985.

Table 1.10

Cloud cover	*Percentage of month*
Stormy	12
Heavy	20
Average	26
Light	32
None	10

Represent this information in a pie chart.

Exercise 1.5

The weekly take-home pay of a group of 1000 people chosen at random for January 1985 is given in Table 1.11.

Table 1.11

Pay (£x)	*Frequency*
$x \leqslant 60$	41
$60 < x \leqslant 85$	43
$85 < x \leqslant 90$	51
$90 < x \leqslant 100$	142
$100 < x \leqslant 110$	231
$110 < x \leqslant 120$	212
$120 < x \leqslant 140$	175
$140 < x \leqslant 160$	105

(a) Illustrate this information on a cumulative frequency graph.
(b) Calculate the proportion of people who take home less than £70 per week.
(c) Draw a histogram to show the frequency distribution.

Exercise 1.6

An examination paper was taken by one hundred candidates who scored, out of a total of 100 marks, the marks shown in Table 1.12.

(i) Draw a histogram to illustrate the distribution of these marks.
(ii) Draw a cumulative frequency polygon.

Table 1.12 Marks scored in an examination

Mark range	*Frequency*
< 30	2
30-39	10
40-49	16
50-59	28
60-69	30
70-79	10
80 and over	4

Exercise 1.7

Table 1.13 shows the number of females, in various age groups, who married for the first time in a certain district in 1979. The age recorded for each female is her age last birthday.

Table 1.13

Age	16-	21-	26-	31-	36-	46-	56-	71-
Number of females	42	113	85	53	27	18	12	0

Construct the cumulative frequency table and draw the cumulative frequency curve.

Exercise 1.8

Table 1.14 shows the marks, collected into groups, of 400 candidates in an examination. The maximum mark was 99.

Table 1.14

Marks	0-9	10-19	20-29	30-39	40-49	50-59	60-69	70-79	80-89	90-99
Number of candidates	10	26	42	66	83	71	52	30	14	6

Compile the cumulative frequency table and draw the cumulative frequency curve.

If the minimum mark for Grade A was fixed at 74, estimate from your curve the percentage of candidates obtaining Grade A.

Exercise 1.9

The masses at birth of 250 babies born in a certain hospital were recorded in kg to the nearest 0.1 kg. Table 1.15 shows the frequency distribution of these masses.

Table 1.15

Mass(kg)	1.0-1.4	1.5-1.9	2.0-2.4	2.5-2.9	3.0-3.4	3.5-3.9	4.0-4.4
Frequency	2	6	17	60	95	60	10

State the limits between which the masses of the babies in the group 2.5–2.9 must lie.

Construct the cumulative frequency table, and draw the cumulative frequency curve. (UCLES)

Exercise 1.10

Table 1.16 shows the times, measured to the nearest 0.1 second, taken for pellets to dissolve in water. Draw a cumulative frequency diagram or ogive for the data. Estimate the mode and the limits for the open intervals.

Use the data and the estimates obtained to draw a frequency diagram of the data and mark on the time axis the ninety-five percentile range.

Table 1.16

Time (seconds)	*Number of pellets*
under 20	45
20 and under 21	150
21 and under 22	290
22 and under 22.5	208
22.5 and under 23	295
23 and under 23.5	330
23.5 and under 24	215
24 and under 24.5	105
24.5 and under 25	55
25 and over	14

(L)

Exercise 1.11

Using the data in Table 1.17 draw a diagram for each age group to illustrate the distribution of arrival times of customers at a men's shop.

Illustrate as a histogram in each case the distribution of ages and the distribution of arrival times of the 424 customers to the shop.

Table 1.17 The number of customers who enter a men's shop by age and time of arrival

Age last birthday	*Time* 9 am -10 am	10 am -11 am	11 am -noon	noon -1 pm	1 pm -2 pm	2 pm -3 pm	3 pm -4 pm	4 pm -5 pm	*Average expenditure per customer £*
61 and over	–	6	6	2	1	3	6	–	6.42
41-60	15	30	26	10	16	20	26	9	12.56
31-40	10	9	9	13	8	13	11	14	13.50
25-30	7	5	6	6	11	4	6	14	11.24
21-24	6	4	5	10	8	3	3	3	10.26
15-20	4	1	–	8	12	4	3	8	8.20
Under 15	2	5	4	2	1	6	–	–	1.59

(L)

2 Averages and Measures of Spread

2.1 Summation Notation

The symbol Σ in mathematics means 'sum'.

$\sum_{i=1}^{n} x_i$ means sum the values of x_i with i increasing by 1 and going from 1 to n.

Hence $\sum_{i=1}^{n} x_i = x_1 + x_2 + x_3 + \ldots + x_n$ [If no confusion arises, this can be written Σx_i or Σx]

2.2 The Arithmetic Mean

The *arithmetic mean* of n numbers $x_1, x_2, x_3, \ldots, x_n$ is often denoted by $\overline{x}$. It is defined as

$$\overline{x} = \frac{x_1 + x_2 + \ldots + x_n}{n} = \frac{\sum_{i=1}^{n} x_i}{n}$$

If the value x_1 occurs with frequency f_1 and x_2 with frequency f_2, etc., then

$$\overline{x} = \frac{f_1 x_1 + f_2 x_2 + f_3 x_3 + \ldots f_n x_n}{f_1 + f_2 + \ldots + f_n} = \frac{\sum_{i=1}^{n} f_i x_i}{\sum_{i=1}^{n} f_i}$$

Assumed Mean

If a guess is made at the mean, say A, and the difference between the guess A and x_i is d_i, then

$$\overline{x} = A + \frac{\Sigma d_i}{n}, \text{ where } d_i = x_i - A$$

or

$$\overline{x} = A + \frac{\Sigma f_i d_i}{n}$$

Worked Example 2.1

Use a suitable guess to find the arithmetic mean of the numbers 485, 513, 498, 501, 512, 497.

Solution 2.1

Let A = 500

The values of d_i are $-15, 13, -2, 1, 12, -3$

$$\Sigma d_i = -15 + 13 - 2 + 1 + 12 - 3 = 6$$

$$\Rightarrow \bar{x} = 500 + \frac{6}{6} = 501$$

The value of this technique does not become apparent until larger groups of number are handled.

2.3 Arithmetic Mean for Grouped Data

If the data has been grouped into classes, and the mid-points of the intervals are denoted by x_i, then

$$\bar{x} = A + \frac{\Sigma f_i d_i}{n}$$

where f_i is the frequency associated with x_i, and d_i is the difference between A and x_i.

Note that A must be one of the x_i values. Consider the table of the distribution for the seedlings given in table 1.4. This has been entered with the values of d_i in table 2.1.

Table 2.1

Class	x_i	f_i	$x_i - A = d_i$	$f_i d_i$
3.75-4.24	3.995	4	−1	−4
4.25-4.74	4.495	5	−0.5	−2.5
4.75-5.24	4.995	4	0	0
5.25-5.74	5.495	3	0.5	1.5
5.75-6.24	5.995	3	1	3
6.25-6.75	6.495	1	1.5	1.5
	Total	20	Total	−0.5

In this case $A = 4.995$.

Choose A so that the distribution is roughly evenly spread on either side of A

$$\Rightarrow \bar{x} = 4.995 + \frac{-0.5}{20} = 4.97$$

The column d_i can often be divided by the class width c to make the working even easier. This then becomes known as the *full coding method*, and is illustrated in Worked Example 2.9.

2.4 Weighted Average

On many occasions the variables $x_1, x_2, x_3, \ldots$ have a weighting factor w_1, w_2, w_3 associated with them. In this case the mean is given by

$$\bar{x} = \frac{w_1 x_1 + w_2 x_2 + \ldots}{w_1 + w_2 + w_3 + \ldots}$$

Worked Example 2.2

In a continuously assessed examination course, the marks on 3 different parts carry 20%, 30% and 50% of the total marks respectively. Jane scored 64%, 37% and 41% on these parts, calculate her final mark.

Solution 2.2

$$\text{Final mark} = \frac{20 \times 64 + 30 \times 37 + 50 \times 41}{20 + 30 + 50}$$

$$= 44.4\%$$

2.5 The Median

The *median* of a set of numbers is the middle value or arithmetic mean of the two middle values, if they are arranged in increasing order of magnitude.

Worked Example 2.3

Find the median of the set of numbers (i) {3, 0, 4, 8, −4, 6, 12, 5, 7}; (ii) {0, 0, 4, −3, 8, 12, 14, −3, 7, 26}.

Solution 2.3

(i) In order of magnitude −4, 0, 3, 4, 5, 6, 7, 8, 12
The median is 5 ↑
(ii) In order of magnitude −3, −3, 0, 0, 4, 7, 8, 12, 14, 26
↑ ↑
The median is $\frac{4 + 7}{2} = 5.5$

A set of ordered numbers such as this is called an *array*.

The median on a histogram or frequency curve is the value of x corresponding to the vertical line that divides the area of the histogram or area under the frequency curve into two equal parts.

2.6 Median for Grouped Data

The median for grouped data can be found either by interpolation or by formula.

Consider the results in Table 1.4. Since there are 20 values, the median will be the average of the 10th and 11th values, i.e. the 10.5th value. Hence the median lies in the 4.75–5.24 class. There are 9 values up to the lower class boundary 4.745. The class 4.75–5.24 contains 4 values. The difference between 10.5 and 9 is 1.5. Hence we require the fraction $\frac{1.5}{4}$ of the class width 0.5.

$$\text{The median} = 4.745 + \frac{1.5}{4} \times 0.5$$

$$= 4.9325$$

This method can be generalised to the following formula.

$$\text{Median} = L_1 + \left(\frac{\frac{1}{2}(n+1) - (\Sigma f)_1}{f_{\text{median}}}\right) c$$

[this formula is usually used whether n is odd or even]

where L_1 is the lower boundary of the class containing the median.

n is the total frequency
$(\Sigma f)_1$ is the sum of all the frequencies in the classes below the median class.
f_{median} is the frequency of the class containing the median.
c is the class width of the median class.

Note that the classes do not all have to be of the same width. Referring again to Table 1.4, the median will be the average of the 10th and 11th values. These are in the 4.75–5.24 class.

$L_1 = 4.745, n = 20, (\Sigma f)_1 = 4 + 5 = 9$

$f_{\text{median}} = 4, c = 0.5$

$$\Rightarrow \text{Median} = 4.745 + \left(\frac{10.5 - 9}{4}\right) 0.5$$

$$= 4.9325$$

2.7 The Mode

The *mode* of a set of numbers is that value (or values) which occurs most often.

A set of numbers with one mode is called *unimodal*. A set with two modes is called *bimodal*.

2.8 Modal Class for Grouped Data

For grouped data, the mode is defined by

$$\text{Mode} = L_1 + \left(\frac{\Delta_1}{\Delta_1 + \Delta_2}\right) c$$

L_1 is the lower class boundary of the modal class. Δ_1 is the difference between the frequency of the modal class and the frequency of the next lower class. Δ_2 is the difference between the frequency of the modal class and the frequency of the next higher class. c is the width of the modal class.

Referring to Table 1.4

$L_1 = 4.245, \Delta_1 = 5 - 4 = 1, \Delta_2 = 5 - 4 = 1, c = 0.5$

$$\Rightarrow \text{Mode} = 4.245 + \left(\frac{1}{1+1}\right) 0.5$$

$$= 4.495$$

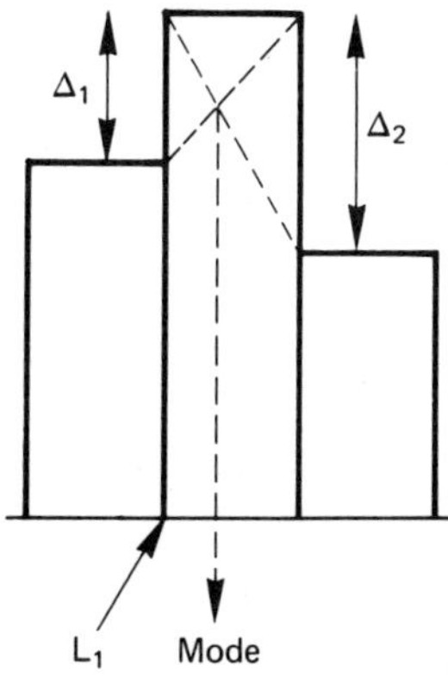

Δ_1 Δ_2 L_1 Mode

Figure 2.1

The geometrical interpretation of the mode of a grouped frequency distribution is shown in Fig. 2.1. If M denotes the mode, then by similar triangles

$$\frac{M - L_1}{c - (M - L_1)} = \frac{\Delta_1}{\Delta_2}$$

$$\Rightarrow (M - L_1)\,\Delta_2 = (c - M + L_1)\,\Delta_1$$

$$\Rightarrow M(\Delta_2 + \Delta_1) = L_1(\Delta_1 + \Delta_2) + c\Delta_1$$

$$\Rightarrow M = L_1 + \frac{c\Delta_1}{\Delta_1 + \Delta_2}$$

2.9 Empirical Relationship between the Mean, Median and Mode

If a frequency curve is moderately skewed, and unimodal, it is approximately true that

Mean – Mode = 3(Mean – Median)

2.10 Geometric Mean

The geometric mean G of $x_1, x_2, x_3, \ldots, x_n$ is defined by

$$G = \sqrt[n]{x_1 x_2 \ldots x_n}$$

The geometric mean is used if proportions are being averaged.

Worked Example 2.4

The total wages paid to management and workers in a small factory for three successive years are shown in Table 2.2.

Table 2.2

Wage bill	*Management*	*Workers*
1983	£61 000	£135 000
1984	£68 000	£160 000
1985	£72 000	£180 000

Find the average proportion of the salary bill paid to management.

Solution 2.4

The proportions for the three years are given by

$$\frac{61\,000}{196\,000} = 0.311, \quad \frac{68\,000}{228\,000} = 0.298, \quad \frac{72\,000}{252\,000} = 0.286$$

$$\text{The average proportion} = \sqrt[3]{0.311 \times 0.298 \times 0.286}$$

$$= 0.298$$

2.11 Harmonic Mean

The harmonic mean H of $x_1, x_2, x_3, \ldots, x_n$ is defined by

$$H = \frac{n}{\sum_{i=1}^{n} \frac{1}{x_i}} = \frac{n}{\frac{1}{x_1} + \frac{1}{x_2} + \ldots + \frac{1}{x_n}}$$

The harmonic mean is used when averaging averages, for example, average speeds.

Worked Example 2.5

The average speeds (km/h) achieved by a motor cyclist in 6 laps of the Isle of Man T.T. race were 201, 198, 185, 200, 187, 202. Find the average speed for the full 6 laps.

Solution 2.5

Since the time for each lap is different, you cannot just find the arithmetic mean. The harmonic mean is appropriate.

$$\text{average speed} = \frac{6}{\frac{1}{201} + \frac{1}{198} + \frac{1}{185} + \frac{1}{200} + \frac{1}{187} + \frac{1}{202}}$$

$$= 195.3 \text{ km/h}$$

2.12 Quartiles, Deciles and Percentiles

If a set of data is arranged in order of magnitude, the values which divide the set into four equal parts, are called the *quartiles*.

The lower quartile is denoted by Q_L ; $[(N + 1)/4]$
The upper quartile is denoted by Q_U . $[3(N + 1)/4]$
The middle quartile is of course the median. $[(N + 1)/2]$

If the data is divided into 10 equal parts, they are called *deciles*, and if it is divided into 100 parts, they are called *percentiles.* $[(N + 1)/100]$

For a grouped frequency table, the quartiles are computed as illustrated in Worked Example 2.6, or measured from a cumulative frequency curve.

Worked Example 2.6

With reference to the information in Table 1.6, find

(i) the lower quartile
(ii) the 55th percentile, of the student ages.

Solution 2.6

(i) There are 97 students, the lower quartile is found at

$$\frac{N + 1}{4} = \frac{98}{4} = 24.5$$

The lower group contains 15 students, hence the proportion of the next group required is

$$\frac{24.5 - 15}{23} = \frac{9.5}{23}$$

Hence $Q_L = 15.95 + \frac{9.5}{23}(1.0) = 16.36$ years [since the width of each class is 1.0 years]

(ii) The 55th percentile is found at $\frac{55(N+1)}{100}$

$$= \frac{55 \times 98}{100} = 53.9$$

The first two groups contain 38 students, hence the

$$\text{55th percentile} = 16.95 + \frac{(53.9 - 38)}{31} \times (1.0)$$

$$= 17.46 \text{ years}$$

2.13 Measures of Spread, Range and Mean Deviation

The *range* of a set of numbers is the difference between the largest and smallest value of the set. The *mean deviation* of a set of numbers $x_1, x_2, \ldots, x_n$ is defined by

$$\text{Mean deviation} = \frac{\sum_{i=1}^{n} |x_i - \bar{x}|}{n}$$

$|x_i - \bar{x}|$ denotes the numerical value of the difference between x_i and $\bar{x}$, i.e. $|7 - 9| = 2$.

The mean deviation of the set $x_1, x_2, \ldots, x_n$ which occurs with frequency $f_1, f_2, \ldots, f_n$ is given by

$$\text{Mean deviation} = \frac{\sum_{i=1}^{n} f_i |x_i - \bar{x}|}{n}$$

Worked Example 2.7

Find the mean deviation of the set of numbers $\{-3, 4, 8, -1, 5\}$.

Solution 2.7

The mean $\bar{x} = \frac{1}{5}(-3 + 4 + 8 - 1 + 5) = 2.6$

$\Rightarrow$ the mean deviation $= \frac{1}{5}(|-3 - 2.6| + |4 - 2.6| + |8 - 2.6| + |-1 - 2.6| + |5 - 2.6|)$

$= \frac{1}{5}(5.6 + 1.4 + 5.4 + 3.6 + 2.4) = 3.68$

2.14 Semi-interquartile Range

In section 2.12, the quartiles were defined.

Semi-interquartile range $= \frac{1}{2}(Q_U - Q_L)$

2.15 Standard Deviation

The standard deviation s of a set of values $\{x_1, x_2, \ldots, x_n\}$ is defined by

$$s = \sqrt{\frac{\sum_{i=1}^{n} (x_i - \bar{x})^2}{n}}$$

For a grouped frequency distribution, the formula becomes

$$s = \sqrt{\frac{\Sigma f_i (x_i - \bar{x})^2}{n}}$$

Worked Example 2.8

Use the mean and standard deviation of the set of numbers {4, 5, 8, 9, 10} to find the mean and standard deviation of the following sets of numbers.

(i) 12, 13, 16, 17, 18
(ii) 32, 40, 64, 72, 80
(iii) 1, 1, 2, 2, 5, 5, 6, 6, 7, 7

Solution 2.8

The purpose of this question is to illustrate the effect on the mean and standard deviation, of constant changes to a set of numbers.
The mean of the original set is 7.2.
The standard deviation is 2.315.

(i) Each number here is 8 greater than the original set

mean = 7.2 + 8 = 15.2

Increasing each number by the same amount does not affect the spread hence standard deviation = 2.315.

(ii) Here, each number is multiplied by 8

mean = 8 x 7.2 = 57.6

The standard deviation will be 8 x 2.315 = 18.52

(iii) Consider {1, 2, 5, 6, 7}. These are all 3 less than the original set. Hence mean = 7.2 – 3 = 4.2

Standard deviation = 2.315.

The set {1, 1, 2, 2, 5, 5, 6, 6, 7, 7} consists of two sets {1, 2, 5, 6, 7} and {1, 2, 5, 6, 7}, hence the mean = 4.2, and the standard deviation is unchanged = 2.315.

2.16 Short Methods for Evaluating the Standard Deviation

The formulae in section 2.15 can be rewritten as follows:

$$s = \sqrt{\frac{\sum_{i=1}^{n} x_i^2}{n} - \left(\frac{\sum_{i=1}^{n} x_i}{n}\right)^2} = \sqrt{\overline{x^2} - (\bar{x})^2}$$

$$s = \sqrt{\frac{\sum_{i=1}^{n} f_i x_i^2}{n} - \left(\frac{\Sigma f_i x_i}{n}\right)^2} = \sqrt{\overline{fx^2} - (\overline{fx})^2}$$

If the information is placed into groups, and A is the assumed mean

let $d_i = x_i - A$

also $u_i = \dfrac{d_i}{c}$

We are extending the ideas developed in section 2.3.
Then

$$s = c\sqrt{\frac{\sum_{i=1}^{n} f_i u_i^2}{n} - \left(\frac{\Sigma f_i u_i}{n}\right)^2}$$

This will be referred to as the *full coding method*.

Worked Example 2.9

Table 2.3 gives the daily takings of a small shop during 1985.

Table 2.3

Daily takings (£)	*Number of days*
0.00– 49.99	22
50.00– 99.99	28
100.00–149.99	55
150.00–199.99	53
200.00–249.99	100
250.00–299.99	102

Use the full coding method to estimate

(i) the mean daily takings;
(ii) the standard deviation.

Solution 2.9

Table 2.4

Class	x_i	$d_i = x_i - A$	$u_i = \dfrac{x_i - A}{c}$	f_i	$f_i u_i$	$f_i u_i^2$
0 - 49.99	24.995	– 150	– 3	22	– 66	198
50.00– 99.99	74.995	– 100	– 2	28	– 56	112
100.00–149.99	124.995	–50	– 1	55	– 55	55
150.00–199.99	174.995	0	0	53	0	0
200.00–249.99	224.995	50	1	100	100	100
250.99–299.99	274.995	100	2	102	204	408
			Totals	360	127	873

Here A = 174.995, class width c = 50 (Table 2.4).

$$\text{(i) Mean} = £174.995 + \left(\frac{\Sigma f_i u_i}{n}\right)c$$

$$= £174.995 + \frac{127 \times 50}{360}$$

$$= £192.63$$

$$\text{(ii) Standard deviation} = c\sqrt{\frac{\Sigma f_i u_i^2}{n} - \left(\frac{\Sigma f_i u_i}{n}\right)^2} = £50\sqrt{\frac{873}{360} - \left(\frac{127}{360}\right)^2}$$

$$= £75.85$$

2.17 Variance

In many situations in statistics, it is the square of the standard deviation, s^2, which is important. We call s^2 the *variance*.

2.18 Sheppard's Correction for the Variance

When a continuous frequency distribution is classified into groups, small errors are introduced. These tend to cancel out when calculating $\overline{x}$, but because s^2 involves $(x - \overline{x})^2$, the errors here are cumulative. There are certain situations where a correction factor is applied, developed by W. F. Sheppard.

$$\text{Corrected Variance} = \text{Calculated variance} - \frac{1}{12}c^2$$

(c is the class width)

The correction should not be applied unless the following conditions hold.

(i) The distribution is nearly symmetrical.
(ii) The distribution is continuous.
(iii) The distribution goes to zero at either end.
(iv) The class width c is not too large.

Worked Example 2.10

The arithmetic mean and the variance of N values of a variable X are $\overline{x}$ and s^2 respectively. If $z = ax + b$, find the arithmetic mean and the variance of the corresponding values of z.

The set {3, 4, 5, 6, 7} is to be transformed using such a linear transformation so that the mean becomes 25, and the variance 60. Find the values of a and b.

Solution 2.10

$$\overline{z} = \frac{\Sigma ax + b}{N} = \frac{\Sigma ax}{N} + \frac{\Sigma b}{N}$$

$$= \frac{a\Sigma x}{N} + b \;[\text{since } \Sigma b = Nb]$$

$$= a\overline{x} + b$$

$$\text{Var}(z) = \frac{\Sigma z^2}{N} - \left(\frac{\Sigma z}{N}\right)^2$$

$$= \frac{\Sigma (ax+b)^2}{N} - (a\bar{x}+b)^2$$

$$= \frac{\Sigma (a^2x^2 + b^2 + 2abx)}{N} - (a\bar{x}+b)^2$$

$$= a^2 \frac{\Sigma x^2}{N} + \frac{\Sigma b^2}{N} + 2ab\frac{\Sigma x}{N} - (a\bar{x}+b)^2$$

$$= a^2[s^2 + \bar{x}^2] + b^2 + 2ab\bar{x} - a^2\bar{x}^2 - b^2 - 2ab\bar{x}$$

$$= a^2 s^2$$

For the set $\{3, 4, 5, 6, 7\}$, (follows from section 2.16 since $s^2 = \overline{x^2} - \bar{x}^2$)

$$\bar{x} = \frac{3+4+5+6+7}{5} = 5$$

$$s^2 = \frac{(3^2+4^2+5^2+6^2+7^2)}{5} - 25 = 2$$

$\Rightarrow$ For the transformed set,

$60 = a^2s^2 = 2a^2 \Rightarrow a^2 = 30$
$\Rightarrow a = 5.48$

$25 = a\bar{x} + b = 5a + b$

But $a = 5.48 \Rightarrow b = 25 - 5 \times 5.48 = -2.39$
The transformation is $z = 5.48x - 2.39$

2.19 Index Numbers

Changes in a variable or a group of variables with respect to time can be expressed in the form of *index numbers*. Most computed indexes are of a socio-economic nature such as the cost of living index and the Financial Times Industrial Ordinary Share Index, but all require some comparison between a *price* or a *quantity* in a current period to a corresponding price or quantity in a base period. An index number can be multiplied by 100 to give a percentage. This may or may not be written with a percentage sign.

In the following formulae for index numbers we shall use:

p_0 – the price of a unit of one commodity in the base period;
p_1 – the price of a unit of one commodity in the current period;
q_0 – the quantity of a commodity consumed in the base period;
q_1 – the quantity of a commodity consumed in the current period;
w – the weighting applied to a commodity.

(i) Simple Price Index. This is often known as a *price relative* and by definition, Price relative $= \frac{p_1}{p_0}$

(ii) Simple Aggregate Price Index

$$= \frac{\Sigma p_1}{\Sigma p_0}$$

(iii) Weighted Price Relative Index

$$= \frac{\Sigma\left(\frac{p_1}{p_0}\right)w}{\Sigma w}$$

(iv) Base Weighted Aggregate Price Index or *Laspeyres Index*. This is $\frac{\Sigma p_1 q_0}{\Sigma p_0 q_0}$

(v) Current Weighted Aggregate Price Index or *Paasche Index*. This is $\frac{\Sigma p_1 q_1}{\Sigma p_0 q_1}$

Worked Example 2.11

In 1984 the price of three commodities A, B and C were 80p, 65p and 55p respectively. Given that the prices of the commodities in 1985 were 97p, x p and 63p and the simple aggregate price index was 114, find x.

Solution 2.11

$\frac{\Sigma p_1}{\Sigma p_0} = 1.14$ [Note: 114 has been divided by 100]

$$\Rightarrow \frac{97 + x + 63}{80 + 65 + 55} = 1.14$$

$\Rightarrow x + 160 = 1.14 \times 200$

$\Rightarrow x = 68$

Worked Example 2.12

A family's budget is influenced mainly by four commodities. The price per unit is shown for the years 1982 and 1983.

Commodity	*Weighting*	*Price/unit 1982*	*Price/unit 1983*
A	3	0.25	0.29
B	1	0.80	0.70
C	2	0.64	0.68
D	4	1.25	1.30

Calculate a weighted price relative index for 1983, using 1982 as base. (SUJB)

Solution 2.12

$$\text{Weighted price relative index} = \frac{\Sigma\left(\frac{p_1}{p_0}\right)w}{\Sigma w}$$

$$= \frac{\left(\frac{0.29}{0.25}\right) \times 3 + \left(\frac{0.7}{0.8}\right) \times 1 + \left(\frac{0.68}{0.64}\right) \times 2 + \left(\frac{1.3}{1.25}\right) \times 4}{10}$$

$$= 1.064$$

This could be stated as $1.064 \times 100 = 106.4\%$

Worked Example 2.13

Table 2.5 gives the consumption and prices of typing and duplicating paper used in a school office for the years 1980 and 1985.
Calculate (i) the Laspeyres Index.
(ii) the Paasche Index.

Table 2.5

	Prices/ream		*Quantities (reams)*	
	1980	*1985*	*1980*	*1985*
Typing	£4.85	£6.15	88	120
Duplicating	£3.80	£4.08	460	580

Solution 2.13

(i) Laspeyres Index $= \dfrac{6.15 \times 88 + 4.08 \times 460}{4.85 \times 88 + 3.80 \times 460}$

$= \dfrac{2418}{2174.8} = 1.112$

(ii) Paasche Index $= \dfrac{6.15 \times 120 + 4.08 \times 580}{4.85 \times 120 + 3.80 \times 580}$

$= \dfrac{3104.4}{2786} = 1.114$

(vi) Quantity Index Numbers are calculated in a similar way to price indexes but with an interchange of the values p and q.

(vii) Comparison of Index Numbers. The simple aggregate price index makes no allowances for the differences in relative importance for different items within the index and in particular it can be affected by the prices of different commodities being measured in different quantities. The weighted price relative index overcomes this problem to some extent, but as the weightings are usually proportional to the amounts we spend on commodities, no account is taken of any change in spending patterns.

With rapid price rises, the weights in the calculation of Laspeyres index become outdated. As prices rise so consumption tends to fall and this index tends to overstate the increase in prices. This is not always the case, however, as was shown in Worked Example 2.13. However, a Paasche index uses current quantities and is therefore up-to-date but by the same argument it tends to understate the increase in prices.

2.20 Time Series (Moving Averages)

The variations of a quantity with respect to time can be illustrated using moving averages. The method is illustrated by the following example.

Worked Example 2.14

The quarterly sales of a magazine over a three-year period are given in Table 2.6. Plot a graph of these figures. On the same graph, plot the four-point moving averages. Fit the trend line, and predict the sales for the first quarter of 1986.

Table 2.6

	Magazine sales (1000's)		
	1983	*1984*	*1985*
First quarter	64	68	68
Second quarter	66	70	73
Third quarter	71	78	80
Fourth quarter	70	71	72

Solution 2.14

The purpose of this type of graph is to smooth out seasonal variations. Since a four-point average has been asked for, the average of results 1–4 is found, and then 2–5, etc. These are as follows:

Sales	64	66	71	70	68	70	78	71	68	73	80	72
Moving Averages		67.75	68.75	69.75	71.5	71.75	71.75	72.5	73	73.25		

The results are illustrated in Fig. 2.2. The moving average points are normally plotted *mid-way* between the points being averaged.

The *trend line* is the line of best fit (straight) for the moving average points.

The deviation for the first quarters d_1, d_2, d_3 from the best fit line are approximately $-3, -2, -5$.

The average is $\frac{1}{3}(-3 - 2 - 5) = -3.3$.

The trend point for the first quarter of 1986 is 75.5.

Hence predicted sales = 75.5 – 3.3 = 72.2 thousand,

i.e. Total sales = 72 200.

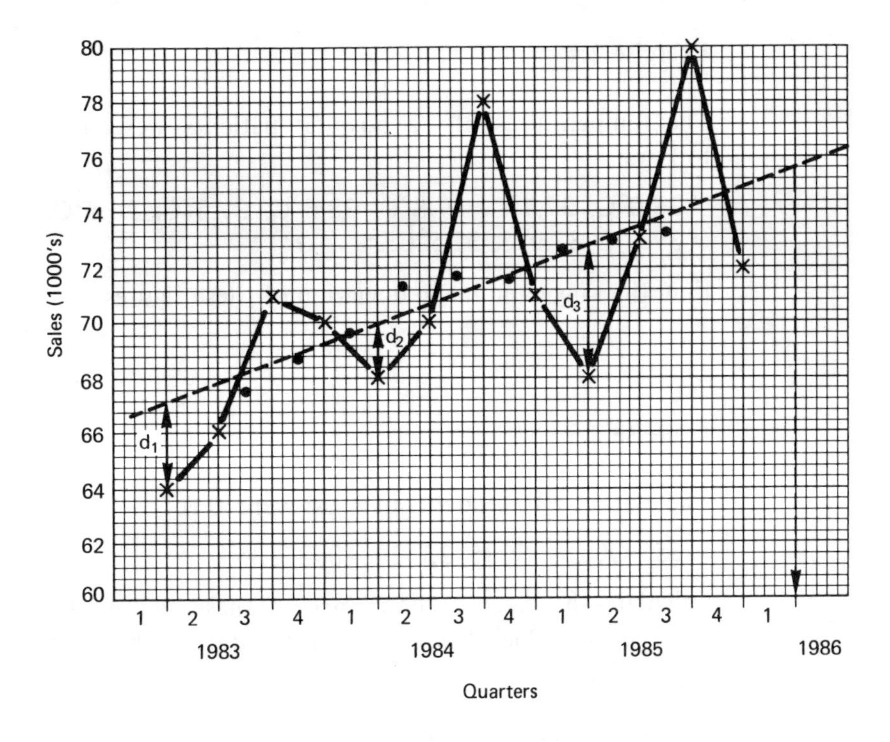

Figure 2.2

2.21 Exercises

Exercise 2.1

With reference to Table 1.12:

(i) Estimate the mean mark, explaining the limitation of your calculation.
(ii) Estimate the median mark and the interquartile range.
(iii) State whether you consider the mean or the median to be the better measure of the average performance of these candidates, giving reasons for your answer.

Exercise 2.2

Use the graph from Exercise 1.7 to find:

(i) The median age at marriage of these females;
(ii) The percentage of these females not married by the age of 40.

Exercise 2.3

Use the graph from Exercise 1.9 to find:

(i) The 60th percentile;
(ii) The interquartile range;
(iii) The percentage of babies whose masses lie between 2.4 kg and 3.6 kg.

Exercise 2.4

The temperatures of 12 hospital patients were measured. The results were as follows:

38.0°C, 37.7°C, 37.8°C, 38.6°C, 37.0°C, 36.8°C, 37.4°C, 38.8°C, 38.7°C, 37.8°C, 37.9°C, 37.5°C.

Find the range and mean deviation of these figures.

Exercise 2.5

The marks of 10 students in a test were as follows:

35, 42, 28, 37, 35, 33, 29, 40, 36, 35.

Calculate the mean and variance of these figures. In order to make the marks more 'presentable' they are transformed using the formula:

$y = 2x - 10$

where y is the new mark and x is the original mark. Calculate the new mean and variance and comment on any relationship between the old and new variances.

Exercise 2.6

A survey of 500 households gave the data in Table 2.7 for distribution of weekly income.

Table 2.7

Income (£)	*Number of households*
40.00- 69.99	30
70.00- 99.99	120
100.00-129.99	140
130.00-159.99	110
160.00-189.99	80
190.00-219.99	20

(a) Use the full coding method, with arbitrary origin at the middle of the third class interval, to obtain:
(i) the mean; (ii) the standard deviation.

(b) Find the median.

Exercise 2.7

One hundred and fourteen teams started the Four Inns Competitive Walk in 1988, and Table 2.8 shows the number of complete teams who reached the nth check point. The first check point is at the start of the walk, and the 12th check point is at the end of the walk.

Table 2.8

n	1	2	3	4	5	6	7	8	9	10	11	12
No. of teams	114	110	109	109	104	103	93	84	69	66	62	60

A team is deemed to have retired at the nth check point if the complete team reaches the nth check point, but not the $(n + 1)$th check point.

Draw up a table showing the number of teams retiring at the nth check point, for $n = 1, 2, \ldots, 11$, and display your tabled information in a diagram.

The times taken by the 60 teams which completed the course are summarised in Table 2.9.

Table 2.9

Time, t, in hours	$7 \leqslant t < 9$	$9 \leqslant t < 11$	$11 \leqslant t < 13$	$13 \leqslant t < 15$	$15 \leqslant t < 17$	$17 \leqslant t < 19$	$19 \leqslant t < 21$
No. of teams	2	11	13	17	11	5	1

Calculate, showing your method:
(i) an estimate of the mean of t, the time taken to complete the course.
(ii) an estimate of the standard deviation of t. (COSSEC)

Exercise 2.8

(a) Six students obtained marks of 20, 19, 25, 29, x and $2x$. Given that the mean of these marks is 24, calculate, without using a calculator, the value of x and the standard deviation of the marks.

(b) A group of readings has a mean of 32 and a standard deviation of 8. Write down the mean and the standard deviation of the group of readings obtained when
(i) each of the original readings is multiplied by 3,
(ii) each of the original readings is halved and then decreased by 6.

(c) Three examination papers are each marked out of 100. These raw marks are weighted and added to give a total which is then scaled down to produce a composite mark out of 100. Table 2.10 shows the weights used and the raw marks of three students.

Table 2.10

	Paper A	*Paper B*	*Paper C*
Weight	1	3	2
John	58	72	46
Peter	74	58	62
Helen	60	68	x

(i) Find a composite mark for John and for Peter and hence determine which of these two students performed better on the three papers.
(ii) Given that Helen obtained a composite mark of 70, find the value of x. (UCLES)

Exercise 2.9

(a) Given the set $\{a, b, c, d, e\}$ with mean m and standard deviation s, what are:

(i) the mean and standard deviation of the set $\{a + k, b + k, c + k, d + k, e + k\}$;
(ii) the mean and standard deviation of the set $\{ak, bk, ck, dk, ek\}$?

In a test, 5 candidates obtained marks of 6, 8, 10, 14, 17. Calculate the mean and standard deviation of these marks. (Your method must be clearly shown).
It is necessary to scale these marks so that they have a mean of 60 and a standard deviation of 20. Calculate the new marks.

(b) Two sets, each with 20 members, have the same standard deviation of 5. The first set has a mean of 17, and the second a mean of 22. Find the standard deviation of the set obtained by combining the given 2 sets. (SUJB)

[Hint: see Exercise 2.11]

Exercise 2.10

Draw suitable statistical diagrams to illustrate the distribution given in Table 2.11. Obtain appropriate estimates of the mean, the median and the mode, and mark these on your diagrams. Explain why these measures of location differ in this distribution. In what circumstances would you expect them all to be the same?

Table 2.11 Hours of sunshine recorded at Bathwell in June 1977

Hours of sunshine per day	*Number of days*
Under 2	2
2 and under 4	7
4 and under 8	10
8 and under 10	6
10 and under 12	4
12 and over	1

(L)

Exercise 2.11

Show, from the basic definition, why the standard deviation of a set of observations $x_1, x_2, x_3 \ldots, x_i, \ldots, x_n$ with mean $\bar{x}$ may be found by evaluating

$$\sqrt{\frac{\Sigma x_i^2}{n} - \bar{x}^2}$$

(a) Find, showing your working clearly and not using any pre-programmed function on your calculator, the standard deviation of the following frequency distribution:

x	25	26	27	28
f	2	0	15	11

(b) The average height of 20 boys is 160 cm, with a standard deviation of 4 cm. The average height of 30 girls is 155 cm, with a standard deviation of 3.5 cm. Find the standard deviation of the whole group of 50 children.

(SUJB)

Hint: $s^2 = \dfrac{n_1 s_1^2 + n_2 s_2^2}{n_1 + n_2} + \dfrac{n_1 n_2}{(n_1 + n_2)^2}(\bar{x}_1 - \bar{x}_2)^2$

Exercise 2.12

In order to estimate the mean length of leaves from a certain tree a sample of 100 leaves was chosen and their lengths measured correct to the nearest mm. A grouped frequency table was set up and the results were as follows:

Mid-interval value (cm)	2.2	2.7	3.2	3.7	4.2	4.7	5.2	5.7	6.2
Frequency	3	5	8	12	18	24	20	8	2

(i) Display the table in the form of a frequency polygon and describe the distribution exhibited by this polygon.
(ii) Calculate estimates for the mean and standard deviation of the leaf lengths using an assumed mean of 4.7 cm.
(iii) What are the boundaries of the interval whose mid-point is 3.7 cm?
(iv) Construct a cumulative frequency table and use it to estimate the sample median.

(SUJB)

Exercise 2.13

Two hundred samples of liquid were analysed for the presence of an impurity, which was measured in parts per million correct to the nearest whole number. The results were as follows:

ppm (X)	50–59	60–69	70–79	80–89	90–99	100–109	110–119	120–129
Frequency	1	6	23	57	51	42	16	4

Using an assumed mean of 84.5 (or otherwise) calculate the mean and standard deviation of X. (SUJB)

Exercise 2.14

(a) The quarterly sales figures for a company over 3 years were:

	1st quarter	*2nd quarter*	*3rd quarter*	*4th quarter*
1976	33 600	36 600	38 600	40 600
1977	36 100	40 000	40 900	42 100
1978	37 900	42 000	44 500	47 600

Plot these figures on a graph, and also the 4-quarter moving average. Predict the sales for the first quarter of 1979.

(b) The results of a survey examining the heights of a sample of 150 people are given in Table 2.12

Table 2.12

Under 5 feet	1
5 feet and under 5 feet 2 inches	4
5 feet 2 inches and under 5 feet 4 inches	14
5 feet 4 inches and under 5 feet 6 inches	23
5 feet 6 inches and under 5 feet 8 inches	35
5 feet 8 inches and under 5 feet 10 inches	38
5 feet 10 inches and under 6 feet	22
6 feet and over	13

Draw the cumulative frequency curve and estimate the median and semi-interquartile range. (SUJB)

Exercise 2.15

Explain briefly the reasons for using moving averages. Table 2.13 shows some of the quarterly trading profits, in thousands of pounds, of a business in three successive years.

Table 2.13

Year	*1st quarter*	*2nd quarter*	*3rd quarter*	*4th quarter*
1	13.2			
2	18.0			
3		39.8	80.6	19.6

(i) Given that the first 4-quarterly moving average is 30.0, calculate the second moving average.
(ii) Given that the graph of the moving averages is a straight line
 (a) calculate the value of the trading profit for the first quarter of year 3;
 (b) estimate the value of the trading profit for the first quarter of year 4.

(UCLES)

Exercise 2.16

(a) The price of a commodity in 1980 was £42. The index number for the price of this commodity in 1970 was 125 based on 1960; in 1980 it was 84 based on 1970. Calculate:

 (i) the index number in 1980 based on 1960;
 (ii) the prices of the commodity in 1960 and 1970;

 It may be assumed that the prices were constant throughout the years in question.

(b) Table 2.14 shows the numbers of cars sold (in thousands correct to the nearest hundred) by a certain manufacturer in the quarters of 1978, 1979 and 1980.

Table 2.14

	Quarter			
	1	*2*	*3*	*4*
1978	59.1	51.2	60.3	49.8
1979	48.3	45.7	50.1	41.2
1980	38.7	36.1	40.4	32.3

Display the data on a graph together with the four-quarterly moving averages. Estimate the number of cars which will be sold in 1981. (SUJB)

Exercise 2.17

The Bitsirusti car company compile data regularly on all aspects of their business. Table 2.15 gives details of employee numbers and average wage rates for three grades for the years 1972, 1976 and 1980.

Table 2.15

	1972		*1976*		*1980*	
	Wage rate (£)	*Numbers*	*Wage rate (£)*	*Numbers*	*Wage rate (£)*	*Numbers*
Line workers	36	200	64	300	119	180
Maintenance fitters	46	40	70	50	105	45
Quality inspectors	55	10	65	8	93	5

Construct and evaluate an appropriate index number for this company to reflect the change in labour costs with respect to these three grades for

(i) 1976 relative to 1972;
(ii) 1980 relative to 1972.

Comment briefly upon the changes that have taken place. (AEB 1982)

3 Probability

3.1 Definitions and Keywords

Consider the following:

(i) The rolling of a die and the score of the top face recorded.
(ii) The throwing of a drawing pin and the position in which it lands recorded.

Each of the above is an example of a *random experiment* or *trial*. The repetition of a trial leads to a set of results or observations but in no instance could we be sure of the *outcome* of a particular trial. Unless of course we had some prior knowledge, i.e. the die was not a fair die but comprised of six 6's!

The *outcome space*, sometimes known as the *possibility space*, or *sample space*, is the set of all possible results from an experiment. This set can be either finite or infinite.

An *event* is any subset of the outcome space. Events which are equally common are called equally likely events.

In (i) above the outcome space S for a fair die can be represented by

$S = \{1, 2, 3, 4, 5, 6\}$

The event E, of throwing a 5 can be represented by

$E = \{5\}$

For a fair die, each of the events $E_i = \{i\}$, where i is a positive integer less than or equal to 6, is equally likely. This symmetrical property makes it reasonable to assume that a *probability* figure assigned to each E_i should be the same. Events of this nature are said to be *equiprobable*.

3.2 Theoretical Definition of Probability

If a trial has a set of equally likely outcomes S, then the probability that an event E occurs is given by $n(E)/n(S)$ (where $n(E)$ and $n(S)$ are the number of elements in the respective sets).

Notation: $$P(E) = \frac{n(E)}{n(S)}$$

(probability of event E occurring)

$E \subseteq S \Rightarrow 0 \leqslant P(E) \leqslant 1$

If S comprises $E_1, E_2, E_3, \ldots, E_n$ disjoint events

i.e. $S = E_1 \cup E_2 \cup \ldots \cup E_n$

and for all i, j $E_i \cap E_j = \phi$

then $\sum_{i=1}^{n} P(E_i) = 1$

Clearly this definition helps us to assign a positive fraction to each event in the rolling of a fair die.

Worked Example 3.1

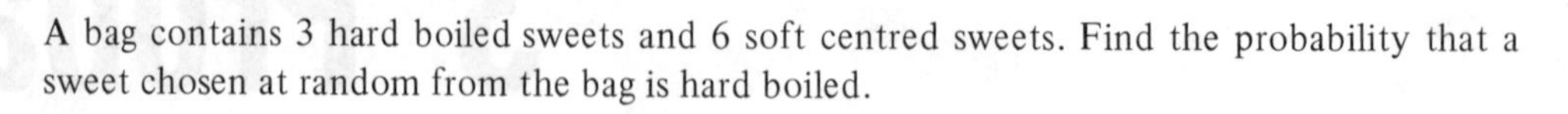

A bag contains 3 hard boiled sweets and 6 soft centred sweets. Find the probability that a sweet chosen at random from the bag is hard boiled.

Solution 3.1

Each of the sweets is equally likely to be chosen and if S represents the set of sweets in the bag and E represents the set of hard boiled sweets,

$n(E) = 3, \qquad n(S) = 9$

Therefore $P(\text{hard boiled sweet chosen}) = \dfrac{3}{9} = \dfrac{1}{3}$

For the drawing pin however the position is not so obvious. In this case we would have to resort to observations and count the *relative frequencies* with which the events:

⊥ – point uppermost

∠ – point touching surface

occur. We would expect the relative frequencies of the two events to approach limits as the number of trials increases. This stability is the basis of the following definition.

3.3 Relative Frequency Definition of Probability

If an event E happens m times in n trials then

$$P(E) = \lim_{n \to \infty} \frac{m}{n}$$

The theoretical definition is the idealised relative frequency of an event happening and intuition and experimentation can confirm that the values derived from the two definitions converge as $n \to \infty$.

3.4 Rules of Probability

(a) Complementary Events

If the probability of an event E happening is $P(E)$ then the probability that it does not happen is $1 - P(E)$. This is known as the complement of E, denoted by $\bar{E}$

so $P(\bar{E}) = 1 - P(E)$

Worked Example 3.2

Show $P(\bar{E}) = 1 - P(E)$

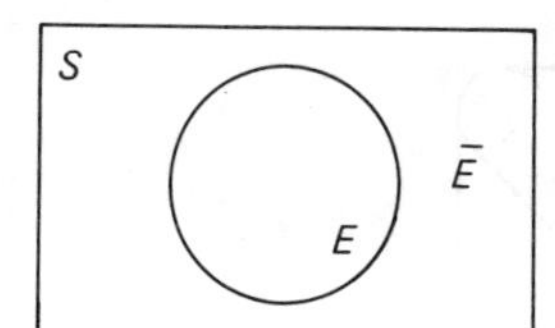

Figure 3.1

Solution 3.2

$$n(\bar{E}) + n(E) = n(S)$$

$$\Rightarrow n(\bar{E}) = n(S) - n(E)$$

Dividing throughout by $n(S)$,

$$P(\bar{E}) = \frac{n(\bar{E})}{n(S)} = 1 - \frac{n(E)}{n(S)}$$

$$= 1 - P(E)$$

(b) Union of Two Events – Addition Rule

The probability of either of two events E_1 and E_2 occurring is given by

$$P(E_1 \cup E_2) = P(E_1) + P(E_2) - P(E_1 \cap E_2)$$

This equation follows immediately from:

$$n(E_1 \cup E_2) = n(E_1) + n(E_2) - n(E_1 \cap E_2)$$

and dividing throughout by $n(S)$. In counting the elements in two sets, those elements in the intersection of the two sets are counted twice hence the subtraction of $n(E_1 \cap E_2)$. It is left to the reader to show that the extension to three sets is:

$$P(E_1 \cup E_2 \cup E_3) = P(E_1) + P(E_2) + P(E_3) - P(E_1 \cap E_2) - P(E_2 \cap E_3) - P(E_3 \cap E_1) + P(E_1 \cap E_2 \cap E_3)$$

Worked example 3.3

Two dice are thrown. What is the probability of scoring either a double, or a sum greater than 8.

Solution 3.3

It is helpful to show a diagram (Fig. 3.2) of the outcome space. Each of the 36 crosses represents an equiprobable event.

$E_1 = \{(1,1), (2,2), (3,3), (4,4), (5,5), (6,6)\}$

$E_2 = \{(3,6), (4,5), (4,6), (5,4), (5,5), (5,6), (6,3), (6,4), (6,5), (6,6)\}$

$n(E_1) = 6, n(E_2) = 10, \; n(E_1 \cap E_2) = 2$

$$P(E_1 \cup E_2) = \frac{6}{36} + \frac{10}{36} - \frac{2}{36} = \frac{14}{36} = \frac{7}{18}$$

N.B. Counting the crosses contained within the two sets in Fig. 3.2 would give the answer immediately.

If E_1 and E_2 are disjoint, i.e. $E_1 \cap E_2 = \phi$

then we say that E_1 and E_2 are *mutually exclusive*. In this instance

$$P(E_1 \cup E_2) = P(E_1) + P(E_2)$$

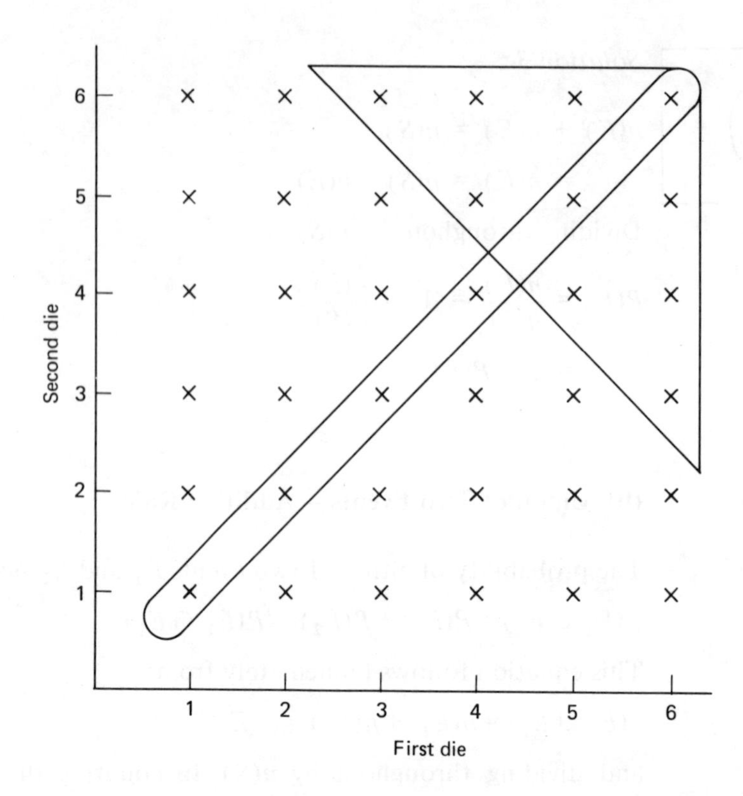

Figure 3.2

(c) Intersection of Two Events – Conditional Probability

If E_1 and E_2 are two events, then the probability that E_2 occurs given that E_1 has occurred is the conditional probability of E_2 *given* E_1 and is defined as:

$$P(E_2/E_1) = \frac{P(E_2 \cap E_1)}{P(E_1)}$$

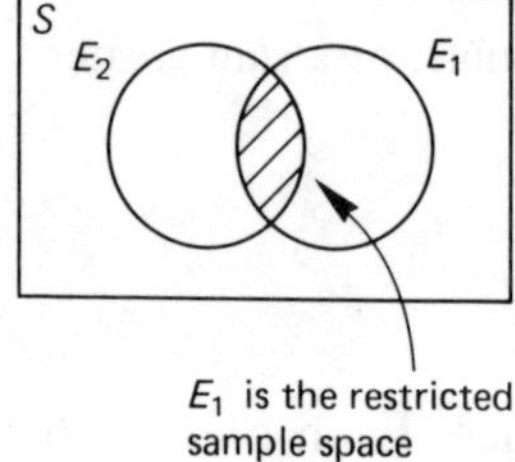

Figure 3.3

From the Venn diagram (Fig. 3.3), the probability of E_2 given that E_1 has occurred is the shaded region divided by the restricted sample space of E_1.

$$\text{So } P(E_2/E_1) = \frac{n(E_2 \cap E_1)}{n(E_1)}$$

$$= \frac{n(E_2 \cap E_1)/n(S)}{n(E_1)/n(S)}$$

$$= \frac{P(E_2 \cap E_1)}{P(E_1)}$$

Worked Example 3.4

Show $P(E_2/E_1)P(E_1) = P(E_1/E_2)P(E_2)$

Solution 3.4

$$P(E_2/E_1)P(E_1) = P(E_2 \cap E_1)$$

and

$$P(E_1/E_2)P(E_2) = P(E_1 \cap E_2)$$

as $E_1 \cap E_2 = E_2 \cap E_1$

$\Rightarrow P(E_2/E_1)P(E_1) = P(E_1/E_2)P(E_2)$

Worked Example 3.5

In Worked Example 3.3, what is the probability of scoring a double given that the total on the two dice is greater than 8?

Solution 3.5

Defining E_1 and E_2 as before:

$P(E_1 \cap E_2) = 2/36, P(E_2) = 10/36$

$$P(E_1/E_2) = \frac{P(E_1 \cap E_2)}{P(E_2)} = \frac{2/36}{10/36} = \frac{1}{5}$$

Worked Example 3.6

It is given that two computers in a batch of six computers are faulty, but it is not known which two. If two machines are selected at random, find the probability that they are both faulty.

Solution 3.6

The possible outcomes can be illustrated in a *tree diagram.* On each branch we can assign the appropriate probability (see Fig. 3.4).

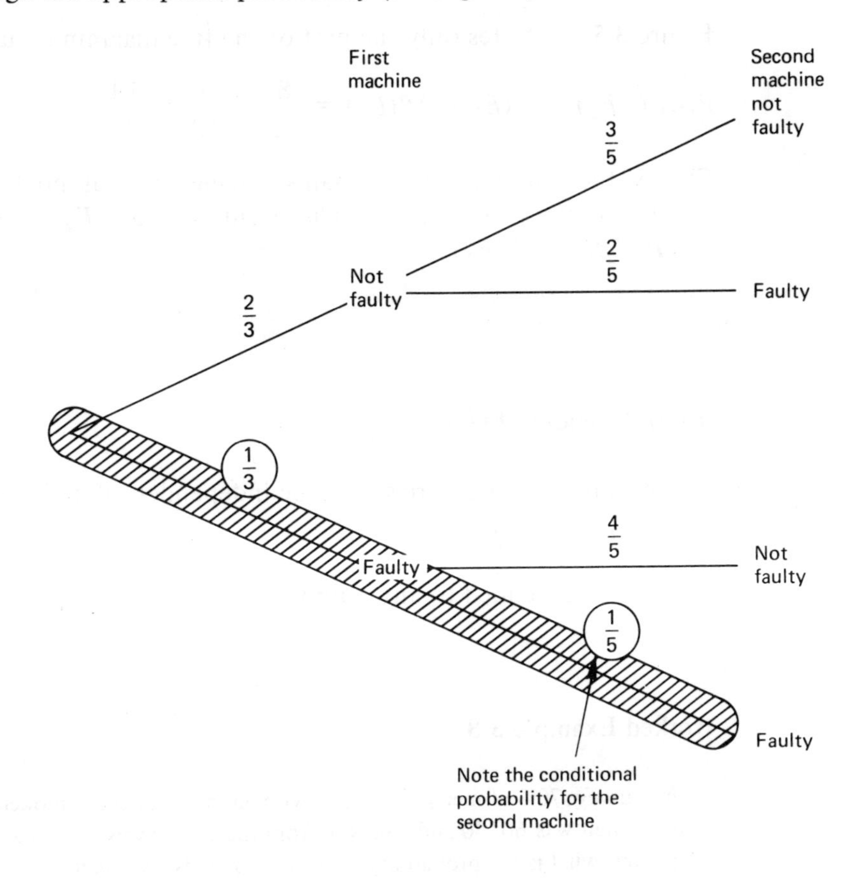

Figure 3.4

If E_1 is the event that the first machine is faulty and E_2 is the event that the second machine is faulty, we require $P(E_1 \cap E_2)$. Using the formula for conditional probability:

$$P(E_1 \cap E_2) = P(E_2/E_1)P(E_1)$$

$$= \frac{1}{5} \times \frac{1}{3} = \frac{1}{15}$$

This 'path' is clearly marked on Fig. 3.4.

Worked Example 3.7

Past experience has shown that a certain examiner at a driving centre passes 70% of men at their first attempt and 80% of women. If a man and a woman taking their test for the first time, are examined by this examiner, what is the probability that they will both pass?

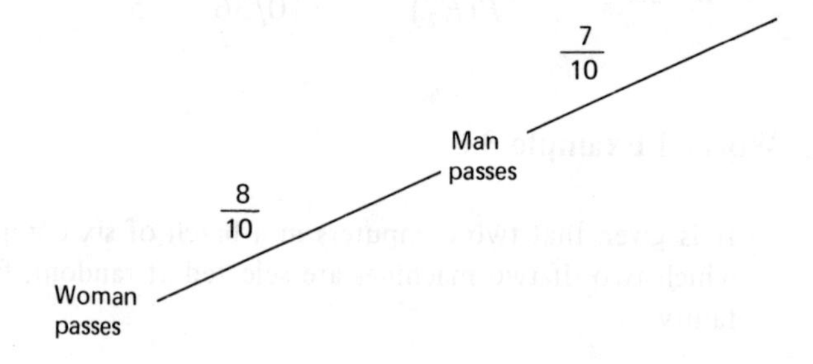

Figure 3.5

Solution 3.7

Figure 3.5 illustrates only the part of the tree diagram required:

$$P(E_1 \cap E_2) = P(E_2/E_1)P(E_1) = \frac{8}{10} \times \frac{7}{10} = \frac{14}{25}$$

Clearly the outcome of the man's driving test has no bearing on the outcome of the woman's driving test. The events E_1 and E_2 are said to be *independent* and $P(E_2/E_1) = P(E_2)$.

(d) Independent Events

Two events E_1 and E_2 are said to be independent if and only if

$$P(E_1 \cap E_2) = P(E_1)P(E_2)$$

This can be extended to any number of events.

Worked Example 3.8

In a survey, 70% of men and 55% of women said that they smoked. If the proportion of men to women was 60:40 and a person from the survey was chosen at random and found to be a smoker, what is the probability that this person is a woman?

Solution 3.8

Let E_M be the event 'a man is chosen'
E_W be the event 'a woman is chosen'
E_S be the event 'a person is a smoker'.

The probability that is required is $P(E_W/E_S)$.

$$\text{Now } P(E_W/E_S) = \frac{P(E_W \cap E_S)}{P(E_S)}$$

Both numerator and denominator can be expressed as follows:

$$P(E_W \cap E_S) = P(E_S/E_W)P(E_W)$$

$$= \frac{11}{20} \times \frac{2}{5} = \frac{22}{100} = \frac{11}{50}$$

$$\text{and } P(E_S) = P(E_S/E_W)P(E_W) + P(E_S/E_M)P(E_M)$$

$$= \frac{11}{20} \times \frac{2}{5} + \frac{7}{10} \times \frac{3}{5} = \frac{32}{50}$$

$$P(E_W/E_S) = \frac{P(E_S/E_W)P(E_W)}{P(E_S/E_W)P(E_W) + P(E_S/E_M)P(E_M)}$$

$$= \frac{11}{50}\Big/\frac{32}{50} = \frac{11}{32}$$

This is an example of *Bayes' Theorem*. Such problems involve finding the probability of a first event happening knowing that a second event has occurred. If the reader finds it difficult to follow the solution above it may be easier to 'see' the solution in the tree diagram (Fig. 3.6). The restricted outcome space is the enclosed two regions of Fig. 3.6, while the event space is the hatched enclosed region.

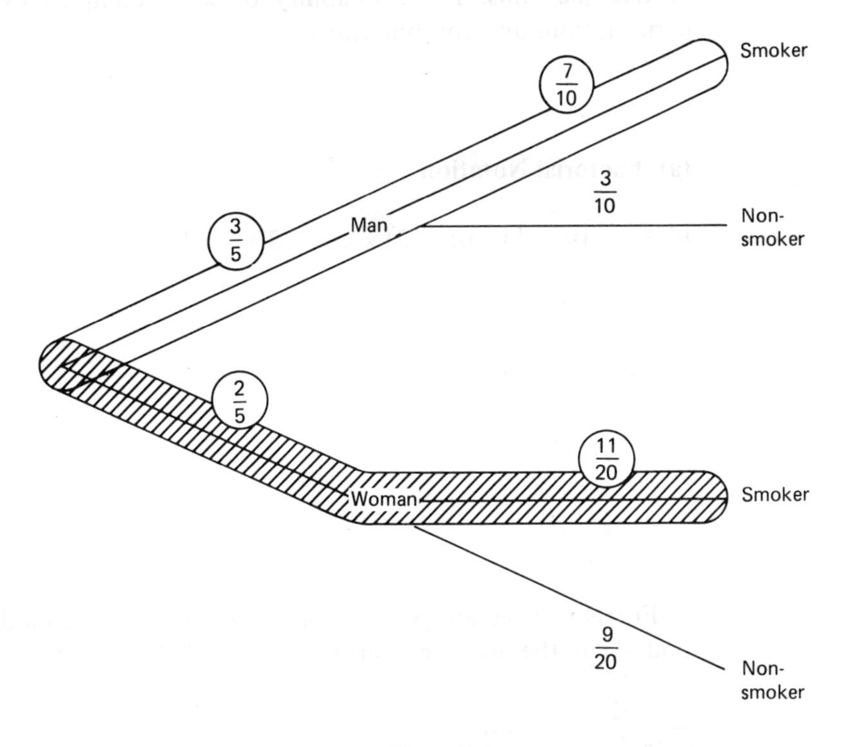

Figure 3.6

Although a general theorem could be stated, the reader is advised always to work from first principles in such problems.

3.5 Infinite Sample Spaces

Worked Example 3.9

A circle of radius r is inscribed inside a square of side $2r$. Find the probability that a point chosen at random is inside the square but outside the circle.

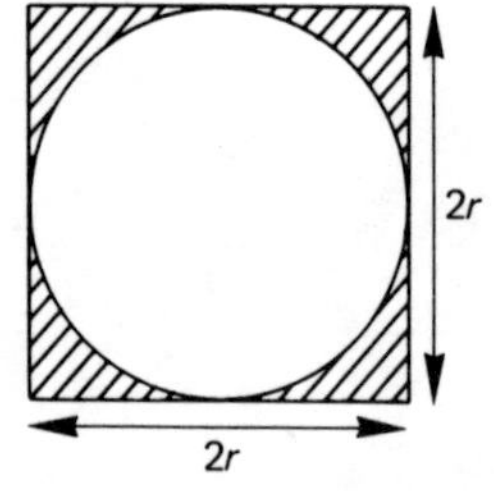

Figure 3.7

Solution 3.9

The probability E, that a point is outside the circle but inside the square, is the ratio of the respective areas; the shaded region and the area of the square (see Fig. 3.7).

$$P(E) = \frac{4r^2 - \pi r^2}{4r^2} = \frac{4 - \pi}{4}$$

3.6 Counting Techniques

We have already seen pictorial representations of counting techniques in the form of tree diagrams. The probability of some complex events require methods of permutations and combinations.

(a) Factorial Notation

$$n! = n \times (n-1) \times (n-2) \times \ldots \times 3 \times 2 \times 1$$

(b) Permutations

Given n distinct objects, the number of arrangements of r of these objects is given by:

$$^nP_r = \frac{n!}{(n-r)!}$$

Further, if we are given n objects, r_1 alike of one kind, r_2 alike of another kind and so on, the number of arrangements of the n objects is

$$\frac{n!}{r_1!r_2!r_3!\ldots} \quad \text{where } \Sigma r_i = n$$

(c) Combinations

Given n distinct objects, the number of selections of r objects without regard to the order is given by:

$$^nC_r = \binom{n}{r} = \frac{n!}{r!(n-r)!}$$

Worked Example 3.10

How many different arrangements are there of the word RUBBER? How many of these arrangements do not have the two B's together?

Solution 3.10

Total number of arrangements = $\dfrac{6!}{2!2!} = 180$

Number of arrangements with the two B's together = $5 \times \dfrac{4!}{2!} = 60$

Number of arrangements without the two B's together = $180 - 60 = 120$

Worked Example 3.11

How many different teams of four people can be chosen from a group of 9 (i) if there are no restrictions and (ii) if one particular person must be included.

Solution 3.11

(i) No restrictions = $\dbinom{9}{4} = \dfrac{9!}{4!5!} = 126$

(ii) if one particular person must be included

$= 1 \times \dbinom{8}{3} = \dfrac{8!}{3!5!} = 56$

Worked Example 3.12

Matt Chovtheday is a striker for Oldcastle United football club. In any match he may score 0, 1, 2, 3 or 4 goals for his side with probabilities 0.2, 0.4, 0.2, 0.1, and 0.1 respectively. Assuming that his performances in successive matches are independent, what is the probability that in three consecutive matches

(i) he scores a total of 10 goals or more,
(ii) he does not score exactly 3 goals in any match,
(iii) his highest score in any match is 3? (AEB 1982)

Solution 3.12

(i) P(He scores a total of 10 goals or more)

$= P(4, 4, 4) + P(4, 4, 3 \text{ in any order}) + P(4, 4, 2 \text{ in any order})$
$\quad + P(4, 3, 3 \text{ in any order})$
$= (0.1)^3 + 3 \times (0.1 \times 0.1 \times 0.1) + 3 \times (0.1 \times 0.1 \times 0.2)$
$\quad + 3 \times (0.1 \times 0.1 \times 0.1)$
$= 0.013$

(ii) P(He does not score exactly 3 goals in any match)

$= (1 - P(\text{He scores 3 goals in a match}))^3$
$= (0.9)^3$
$= 0.729$

(iii) P(Highest score in any match is 3)

He must score at least one 3 but not a 4.

$= P$(He does not score a 4 in a match) –
P(He fails to score more than 2 in a match)
$= (0.9)^3 - (0.8)^3$
$= 0.217$

Worked Example 3.13

Two events A and B are independent. If $P(A \cap B) = 1/3$ and $P(A \cup B) = 5/6$ find possible values for (i) $P(A)$, (ii) $P(B)$.

Solution 3.13

Let $P(A) = x$ and $P(B) = y$

Independence $\Rightarrow P(A \cap B) = P(A)P(B)$

$\Rightarrow \quad 1/3 = xy \qquad$ (i)

and $\quad P(A \cup B) = P(A) + P(B) - P(A \cap B)$

$\Rightarrow \quad 5/6 = x + y - xy \qquad$ (ii)

Substituting (i) into (ii)

$$\frac{5}{6} = \frac{1}{3y} + y - \frac{1}{3}$$

Rearranging, $6y^2 - 7y + 2 = 0$

$\Rightarrow \quad (3y - 2)(2y - 1) = 0$

either $y = 2/3 \Rightarrow x = 1/2$
or $\quad y = 1/2 \Rightarrow x = 2/3$

Worked Example 3.14

Prove that if A and B are independent events, then $\overline{A}$ and $\overline{B}$ are independent events.

Solution 3.14

A and B are independent $\Rightarrow P(A \cap B) = P(A)P(B)$

Now $\quad \overline{A} \cap \overline{B} = \overline{(A \cup B)}$

so

$$
\begin{aligned}
P(\overline{A} \cap \overline{B}) &= P(\overline{[A \cup B]}) \\
&= 1 - P(A \cup B) \\
&= 1 - [P(A) + P(B) - P(A)P(B)] \\
&= 1 - P(A) - P(B) + P(A)P(B) \\
&= [1 - P(A)]\,[1 - P(B)] \\
&= P(\overline{A})P(\overline{B})
\end{aligned}
$$

$\Rightarrow \overline{A}$ and $\overline{B}$ are independent events.

Worked Example 3.15

95% of car drivers wear seat belts, 64% of car drivers involved in serious accidents die if not wearing a seat belt whereas 12% of those that do wear a seat belt die. Calculate, correct to three significant figures, the percentage of drivers involved in serious accidents who died and were not wearing seat belts. (SUJB)

Solution 3.15

Considering the problem in terms of probabilities, it should be seen that it is an example of Bayes' Theorem.

Let E_D be the event 'died as a result of a serious accident'
E_S be the event 'wears a seat belt'

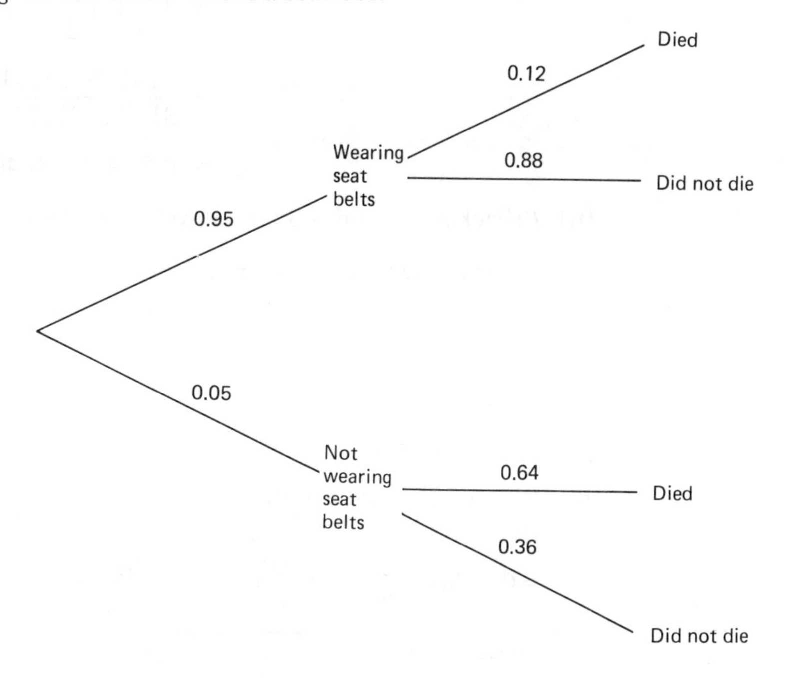

Figure 3.8

$$\text{We require } P(\bar{E}_S/E_D) = \frac{P(E_D/\bar{E}_S)P(\bar{E}_S)}{P(E_D/\bar{E}_S)P(\bar{E}_S) + P(E_D/E_S)P(E_S)}$$

$$= \frac{0.64 \times 0.05}{0.64 \times 0.05 + 0.12 \times 0.95}$$

$$= 0.2192$$

So the proportion to 3 significant figures is 21.9%.

Worked Example 3.16

(i) A box contains 15 tulip bulbs of which 10 will have red flowers and 5 yellow, though it is not possible to determine the colour of the flowers by visual inspection of the bulbs. A package of 5 is to be made up from the 15. If they are selected at random, what is the probability that the package will contain exactly 3 red-flowering bulbs?

(ii) One of the 15 bulbs is found to be damaged and is discarded before the selection is made. Find the conditional probability that the package of 5 contains exactly 3 red-flowering bulbs, given that the damaged bulb is red.

(iii) Hence or otherwise find the probability that the package of 5 contains exactly 3 red-flowering bulbs when one bulb in the original 15 has been thrown away as damaged, given that all bulbs are equally likely to be damaged. (Give all answers to 3 decimal places.) (OLE)

Solution 3.16

(i) Total number of selections $= \binom{15}{5}$

Number of selections which contain exactly 3 red-flowering bulbs

$$= \binom{10}{3}\binom{5}{2}$$

P(Package contains exactly 3 red-flowering bulbs)

$$= \binom{10}{3}\binom{5}{2}\Big/\binom{15}{5}$$

$$= \frac{10!\,.\,5!\,.\,5!\,.\,10!}{3!\,.\,7!\,.\,2!\,.\,3!\,.\,15!}$$

$$= 0.3996 = 0.400$$

(ii) P(Package contains 3 red-flowering bulbs/damaged bulb is red)

$$= \frac{P(\text{Package contains 3 red-flowering bulbs and damaged bulb is red})}{P(\text{Damaged bulb is red})}$$

$$= \binom{9}{3}\binom{5}{2}\Big/\binom{14}{5}$$

$$= 0.4196 = 0.420$$

(iii) P(3 red-flowering bulbs/All bulbs are equally likely to be damaged and thrown)

$$= \underbrace{0.4196}_{\text{(damaged bulb is red)}} \times \frac{2}{3} + \underbrace{\left\{\binom{10}{3}\binom{4}{2}\Big/\binom{14}{5}\right\}}_{\text{(damaged bulb is yellow)}} \times \frac{1}{3}$$

$$= 0.4196 \times \frac{2}{3} + 0.3596 \times \frac{1}{3}$$

$$= 0.400$$

Worked Example 3.17

(a) A game consists of rolling a 2p coin down a ramp onto a flat table divided into 5 cm squares. Given that the diameter of a 2p coin is 26 mm and that the coin is returned if it does not stop on the table, find the probability that the coin stops completely inside a square.

*(b) Find the probability that three points chosen at random on the circumference of a circle lie on a semicircle.

Solution 3.17

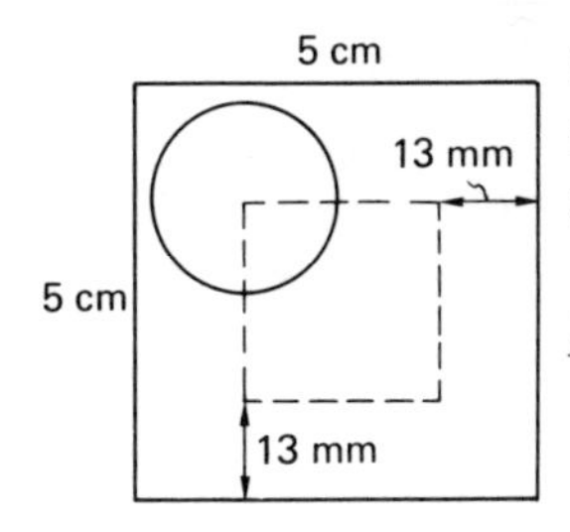

Figure 3.9

(a) In order for the coin to stop completely inside a square, the centre of the coin must be no less than 13 mm from the edge of the square. The centre of the coin should be inside the dotted square.

$$P(\text{Coin stops completely inside a square}) = \frac{\text{area of dotted square}}{\text{area of larger square}}$$

$$= \frac{24 \times 24}{50 \times 50} = \frac{144}{625}$$

*(b) Let the three points be O, X and Y and let the distances OX and OY measured clockwise around the circle be x and y. Now without loss of generality we can choose x such that $x \leqslant C/2$ where C is the length of the circumference. Now in order that the 3 points should lie on a semicircle, there are two possible restrictions on y:

either $y \leqslant C/2$ or $y - x \geqslant C/2$

Figure 3.10 shows diagrammatical representations of the problem. This figure represents the outcome space for x and y and the shaded region is the event space. The required probability is therefore 3/4.

Worked Example 3.18

(a) Three numbers are chosen at random from the set $\{1, 2, 3, \ldots, 10\}$. What is the probability that the smallest of the three numbers chosen is even?

(b) Two boys, A and B, play a game of Beetle; each throws a die in turn and cannot start drawing his Beetle until he obtains a 6. The boy A throws first.

(i) What is the probability that B is the first boy to begin drawing a Beetle, starting on his second throw?

(ii) What is the probability that B is the first to start drawing a Beetle? (SUJB)

Solution 3.18

(a) Number of selections without any restrictions = $\binom{10}{3} = 120$

We require the number of selections in which in turn 2, 4, 6 and 8 are the smallest numbers in each selection. (We are not interested in 10!) So number of selections with these restrictions

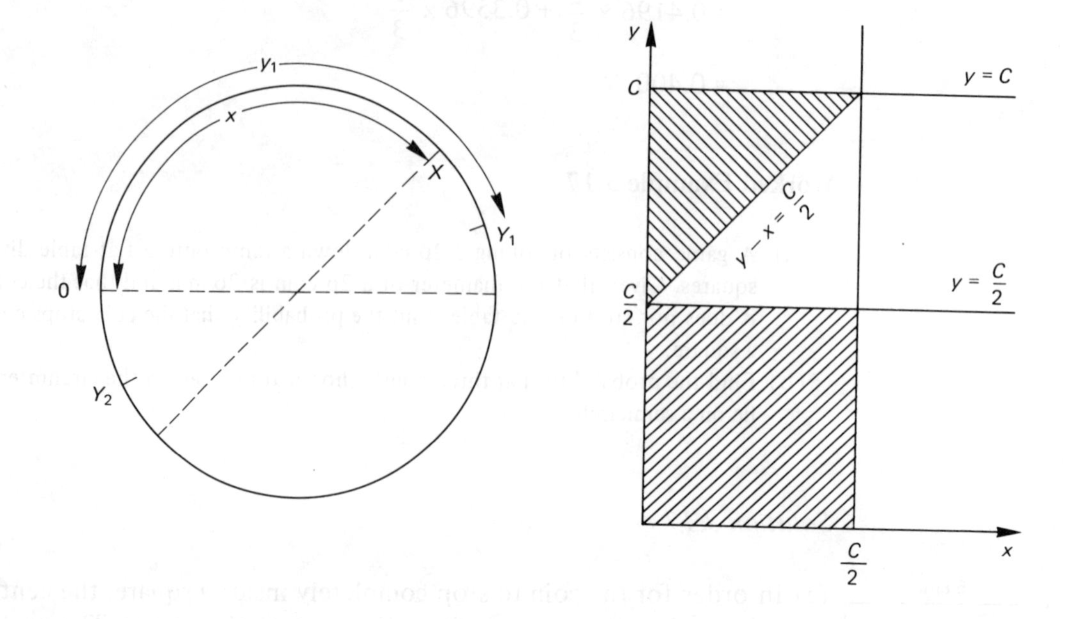

Figure 3.10

$$= \binom{8}{2} + \binom{6}{2} + \binom{4}{2} + \binom{2}{2}$$

$$= \frac{8.7}{2.1} + \frac{6.5}{2.1} + \frac{4.3}{2.1} + \frac{2.1}{2.1}$$

$$= 50$$

So the probability that the smallest of the three numbers is even

$= 50/120 = 5/12$

(b) (i) To answer this we have to appreciate that if B is the first to start on his second throw, he will have failed to start on one attempt already and A will have failed to start twice. The probability of not throwing a 6 is 5/6 and of course is independent of any previous attempts.

P(B is the first to begin drawing a Beetle on his second throw)

A fails B fails A fails B successful

$$= \frac{5}{6} \times \frac{5}{6} \times \frac{5}{6} \times \frac{1}{6} = \frac{125}{1296}$$

(ii) It is theoretically possible for B never to start and in calculating the probability there are an infinite number of terms. If we look at the first few terms we should recognise the pattern of a Geometric Series.

B starts at attempt no.

	1	2	3 ...
Probability	$\frac{5}{6}.\frac{1}{6}$	$\left(\frac{5}{6}\right)^3\left(\frac{1}{6}\right)$	$\left(\frac{5}{6}\right)^5\left(\frac{1}{6}\right)$ ...

The probability required is the sum of this infinite series which has a finite sum as the common ratio is positive and less than 1.

$$P(\text{B wins}) = S_\infty = \frac{5/36}{1 - 25/36} = \frac{5}{11}$$

3.7 Exercises

Exercise 3.1

Two dice are thrown. Find the probability that the scores are consecutive.

Exercise 3.2

Given that $P(A) + P(B) = 1$, $P(A/B) = 2/3$ and $P(B/A) = 3/8$ find $P(A)$.

Exercise 3.3

Given that the proportion of red balls in an urn is 3/4, find the smallest number of red balls if the probability of drawing two balls without replacement and finding only one red ball is 15/38.

Exercise 3.4

How many times should a fair die be thrown if the probability of a one should appear at least once is to be greater than 95%?

Exercise 3.5

The proportion of female students at a college is 55%. If 30% of the male students and 5% of the female students study physics, what is the probability that a physics student chosen at random, is a girl?

Exercise 3.6

Given a straight line ABC, B is the midpoint of the line. A point X is chosen at random along the line. Find the probability that the three lengths AX, XC and AB can form a triangle.

Exercise 3.7

A point is selected at random from inside an isosceles triangle whose sides measure 10, 10 and 16 cm. Find the probability that this point is within 3 cm of a vertex.

Exercise 3.8

How many different arrangements can be made from all of the letters of the word ATTENDED? What is the probability that in an arrangement chosen at random the D's are separated?

*Exercise 3.9

Monsieur R. Andom walks home from work every evening. Figure 3.11 shows the possible routes with six roads running South–North and six running West–East. Given that he always moves either North or East, how many different routes can be take?

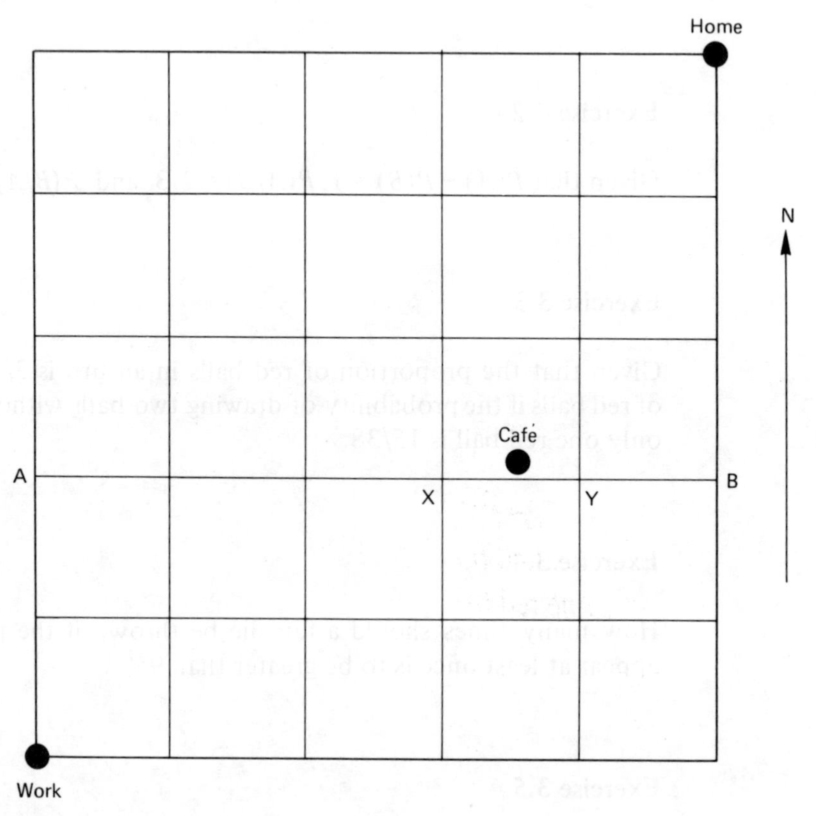

Figure 3.11

The cafe Bernouilli is situated half-way between the two junctions marked X and Y on the West–East road AB. If at each junction, Monsieur Andom chooses a possible route randomly and he never passes the cafe without stopping for a drink, find the probability that on a day that he works, Monsieur Andom can be seen drinking at cafe Bernouilli in the evening.

Exercise 3.10

A box contains twelve balls numbered from 1 to 12. The balls numbered 1 to 5 are red, those numbered 6 to 9 are white, and the remaining three balls are blue. Three balls are to be drawn at random without replacement from the box. Let A denote the event that each number drawn will be even, B the event that no blue ball will be drawn, and C the event that one ball of each colour will be drawn. Calculate (i) $P(A)$, (ii) $P(B)$, (iii) $P(C)$, (iv) $P(A \cap C)$, (v) $P(B \cup C)$, (vi) $P(A \cup B)$.
(WJEC)

Exercise 3.11

(a) In each of the following two games show that the probability that the player concerned will win is $1/2$.
Game 1: The player throws two fair dice together and wins if he throws at least one 5 or if the sum of the two scores is equal to 6 or equal to 7.
Game 2: The player tosses a fair coin four times in succession and wins if he tosses at least two heads successively.
(b) The three events A, B and C are such that A and B are mutually exclusive, A and C are independent, and $P(A) = 0.2, P(B) = 0.1, P(A \cup C) = 0.5, P(B \cup C) = 0.4$.
(i) Evaluate $P(C)$.
(ii) Determine whether B and C are independent, mutually exclusive or neither.
(WJEC)

Exercise 3.12

Out of 50 patients being treated at a clinic for a severe allergy, 10 are chosen at random to receive a new dietary treatment as opposed to the standard drug treatment given to the other 40 patients. It is known that the probability of a cure with the standard treatment is 0.6 whereas the probability of a cure with the new treatment is 0.9. Sometime later, one of these patients returns to the clinic to thank the doctors for his cure. What is the probability that he had been given the new treatment? (AEB 1982)

Exercise 3.13

There are four human blood groups O, A, B and AB and each person must have one of these types of blood. The proportions of the population in these groups are 0.46, 0.40, 0.11 and 0.03 respectively.
(i) If two people are chosen at random, what is the probability that they both have the same blood group?
(ii) If three people are chosen at random, what is the probability that they all have different types?
(iii) If four people are chosen at random, what is the probability that two have group O and the other two have groups which are different from O and different from each other? (SUJB)

Exercise 3.14

Events A, B, C are such that A and C are mutually exclusive and

$P(A) = P(B) = P(C) = x$

$P(A \cap B) = P(B \cap C) = y$

$P(A \cup B \cup C) = 1$

Find an expression for y in terms of x.
Find also the greatest and least possible values of x and find the greatest possible value of $P(\bar{A} \cap B \cap \bar{C})$. (L)

Exercise 3.15

(a) (i) Find the total number of different selections of 4 apples from 12 apples.
(ii) If 3 of these 12 apples are bruised, and a random selection of 4 apples is made, find the probability that the selection will contain precisely one bruised apple.

(b) X and Y are independent events such that $P(X) = 0.7$ and $P(Y) = 0.4$. Calculate (i) $P(X \cup Y)$, (ii) $P[(X \cap Y)/(X \cup Y)]$. (UCLES)

Exercise 3.16

A restaurant offers a lunch menu consisting of 3 soups, 4 entrees and 5 sweets. If, for each meal, a regular diner chooses one soup, one entree and one sweet, how many different meals can he choose? If further, he makes his selection at random, calculate the probability that next week he chooses:

(a) the same meal on Monday and Tuesday,
(b) meals with entirely different courses on Monday and Tuesday,
(c) meals with entirely different courses on Monday, Tuesday and Wednesday. (L)

Exercise 3.17

A and B play a game as follows: an ordinary die is rolled and if a six is obtained then A wins and if a one is obtained then B wins. If neither a six nor a one is obtained then the die is rolled again until a decision can be made. What is the probability that A wins on (i) the first roll, (ii) the second roll, (iii) the rth roll? What is the probability that A wins? (SUJB)

Exercise 3.18

The Vice-Chancellor of the University of Melchester has asked a group of his staff to sit on a committee to consider future developments in the light of the levels of funding promised to all universities. The committee consists of twelve people each representing a different department, and of these, four are professors, namely the professors of Physics, Mathematics, Law and Engineering. At their first meeting all members of the committee sit down at random in a row.

(a) In how many distinct ways may they do this?
(b) What is the probability that the professors of Mathematics and Physics are next to each other?
(c) What is the probability that the professors of Mathematics and Physics are next to each other, but that the professors of Law and Engineering are not?

At their second meeting the professor of Law has arranged for the committee to sit round a circular table. Assuming that all members of the committee still sit down at random, calculate the new answers to questions (a), (b) and (c) above. (AEB 1983)

4 Discrete Probability Distributions

4.1 Discrete Random Variables

In Worked Example 3.3 of the previous chapter we enumerated the sample space of the rolling of two dice in diagrammatical form. In order to find a sum greater than 8, we encircled 10 crosses in the equiprobable space and so

$P(\text{Score greater than } 8) = 10/36$

It is often convenient to see the possible outcomes and the associated probabilities of an event in tabular form and for the rolling of two fair dice the possible scores and probabilities are shown in Table 4.1.

Table 4.1

Score (X)	2	3	4	5	6	7	8	9	10	11	12
Probability	$\frac{1}{36}$	$\frac{2}{36}$	$\frac{3}{36}$	$\frac{4}{36}$	$\frac{5}{36}$	$\frac{6}{36}$	$\frac{5}{36}$	$\frac{4}{36}$	$\frac{3}{36}$	$\frac{2}{36}$	$\frac{1}{36}$

$P(X > 8) = \quad 10/36$

The possible scores, X, are numerical values determined by the outcome of a random experiment. The sample space from which these values are derived is discrete, that is the distribution has a finite number of values, and X is called a *discrete variate* or a *discrete random variable*.

If X is a discrete variable taking values $x_1, x_2, x_3, \ldots, x_n$ and to each x_i there is associated a p_i, such that:

(i) All $p_i \geqslant 0$
(ii) $\Sigma p_i = 1$

then X is a discrete random variable and the set of probabilities forms a *probability function* of X.

Worked Example 4.1

At the local cinema the film showing is 'Discrete Expectations' and three brothers A, B and C will watch this film with probabilities 1/4, 1/3, and 2/5 respectively. The decision a brother makes to watch the film is independent of the other brothers. Find the probability function of the number of brothers who watch the film.

Solution 4.1

Let X be the discrete random variable representing the number of brothers who watch the film and A, B, C represent the respective events that a brother watches the film.

$$P(X = 0) = P(\bar{A} \cap \bar{B} \cap \bar{C}) = P(\bar{A}) . P(\bar{B}) . P(\bar{C})$$

$$= \left(1 - \frac{1}{4}\right)\left(1 - \frac{1}{3}\right)\left(1 - \frac{2}{5}\right) = \frac{3}{10}$$

$$P(X = 1) = P(A \cap \bar{B} \cap \bar{C}) + P(\bar{A} \cap B \cap \bar{C}) + P(\bar{A} \cap \bar{B} \cap C)$$

$$= \frac{1}{4} . \frac{2}{3} . \frac{3}{5} + \frac{3}{4} . \frac{1}{3} . \frac{3}{5} + \frac{3}{4} . \frac{2}{3} . \frac{2}{5} = \frac{9}{20}$$

$$P(X = 2) = P(A \cap B \cap \bar{C}) + P(A \cap \bar{B} \cap C) + P(\bar{A} \cap B \cap C)$$

$$= \frac{1}{4} . \frac{1}{3} . \frac{3}{5} + \frac{1}{4} . \frac{2}{3} . \frac{2}{5} + \frac{3}{4} . \frac{1}{3} . \frac{2}{5} = \frac{13}{60}$$

$$P(X = 3) = P(A \cap B \cap C)$$

$$= \frac{1}{4} . \frac{1}{3} . \frac{2}{5} = \frac{1}{30}$$

In tabular form:

Number of brothers watching film (X)	0	1	2	3
$P(X)$	$\frac{3}{10}$	$\frac{9}{20}$	$\frac{13}{60}$	$\frac{1}{30}$

4.2 Expectation and Variance

The *expectation* of the discrete random variable X, denoted by $E(X)$

$$p_1x_1 + p_2x_2 + p_3x_3 + \ldots + p_nx_n = \Sigma\, p_ix_i$$

In Chapter 2 we saw how the mean of a set of observations was derived from the sum of the products of each x_i with its associated relative frequency f_i/n. The expectation defined above is of the same form with the relative frequencies being replaced by the expected relative frequencies of an event occurring, p_i.

Expectation therefore is synonymous with the mean and the letter μ is often used to represent the mean of a random variable.

Worked Example 4.2

In Worked Example 4.1, find the expectation or the mean number of brothers who watch the film.

Solution 4.2

$$E(X) = \frac{3}{10} \times 0 + \frac{9}{20} \times 1 + \frac{13}{60} \times 2 + \frac{1}{30} \times 3$$

$$= 59/60$$

Worked Example 4.3

The random variable X has the distribution given by

X	1	2	3	4
$P(X)$	$\frac{1}{5}$	$\frac{3}{10}$	$\frac{2}{5}$	$\frac{1}{10}$

Find (i) $E(X)$, (ii) $E(3X)$, (iii) $E(3X + 2)$.

Solution 4.3

(i) $E(X) = \frac{1}{5} \times 1 + \frac{3}{10} \times 2 + \frac{2}{5} \times 3 + \frac{1}{10} \times 4 = \frac{12}{5}$

(ii) Each value of x in the summation for the expectation is replaced by $3x$.

Therefore $E(3X) = \frac{1}{5} \times 3 + \frac{3}{10} \times 6 + \frac{2}{5} \times 9 + \frac{1}{10} \times 12 = \frac{36}{5}$

$= 3E(X)$

(iii) Each x is replaced by $3x + 2$.

Therefore $E(3X + 2) = \frac{1}{5} \times 5 + \frac{3}{10} \times 8 + \frac{2}{5} \times 11 + \frac{1}{10} \times 14 = \frac{46}{5}$

$= 3E(X) + 2$

This example illustrates some of the properties of expectation, namely:

(i) $E(aX) = aE(X)$
(ii) $E(aX + b) = aE(X) + b$

where a and b are constants.

Furthermore (iii) $E(ag(X) + bh(X)) = aE(g(X)) + bE(h(X))$
where g and h are defined functions on X.

Worked Example 4.4

Show that $E[(X - \mu)^2] = E(X^2) - \mu^2$

Solution 4.4

$E[(X - \mu)]^2 = E[X^2 - 2\mu . X + \mu^2]$

using property (iii) above

$= E(X^2) - 2\mu . E(X) + \mu^2$

$= E(X^2) - 2\mu^2 + \mu^2$

$= E(X^2) - \mu^2$

The variance can be defined, using relative frequencies as

$\frac{\Sigma f_i x_i^2}{n} - \left(\frac{\Sigma f_i x_i}{n}\right)^2$, see page 22

$$= \Sigma\left(\frac{f_i}{n}\right) x_i^2 - \left(\Sigma\left(\frac{f_i}{n}\right) x_i\right)^2$$

Replacing f_i/n by probabilities we can define

$$\text{Var}(X) = \Sigma\, p_i x_i^2 - \left(\Sigma p_i x_i\right)^2$$

or

$$\text{Var}(X) = E(X^2) - [E(X)]^2$$

σ^2 is often used to represent the variance of a random variable.

Worked Example 4.5

Calculate Var(X) for Worked Example 4.3.

Solution 4.5

$$\text{Var}(X) = \frac{1}{5} \times 1^2 + \frac{3}{10} \times 2^2 + \frac{2}{5} \times 3^2 + \frac{1}{10} \times 4^2 - \left(\frac{12}{5}\right)^2$$

$$= \frac{33}{5} - \frac{144}{25} = \frac{21}{25}$$

The reader should confirm using this example that

$$\text{Var}(X) = E[(X - \mu)^2]$$

Useful properties of the variance are:

(i) $\text{Var}(aX) = a^2\,\text{Var}(X)$ and

(ii) $\text{Var}(aX + b) = a^2\,\text{Var}(X)$

where a and b are constants.

Worked Example 4.6

Show $\text{Var}(aX) = a^2\text{Var}(X)$

Solution 4.6

$$\begin{aligned}\text{Var}(aX) &= E[(aX)^2] - [E(aX)]^2 \\ &= E(a^2X^2) - [aE(X)]^2 \\ &= a^2E(X^2) - a^2[E(X)]^2 \\ &= a^2[E(X^2) - [E(X)]^2] \\ &= a^2\text{Var}(X)\end{aligned}$$

In Worked Example 4.3, Var(3X) = 9 × Var(X) = 9 × 21/25 = 189/25.

4.3 Joint Distributions

If X and Y are two random variables, we can define a third random variable Z such that $Z = X + Y$.

Worked Example 4.7

Three fair coins are tossed. Let X be the number of Heads thrown and Y be the longest sequence of Heads. Draw up distribution tables for X, Y and Z where $Z = X + Y$ and hence find the mean and variances of the three random variables.

Solution 4.7

The distributions of X and Y are illustrated in Tables 4.2 and 4.3.

Table 4.2

X	0	1	2	3
$P(X)$	$\frac{1}{8}$	$\frac{3}{8}$	$\frac{3}{8}$	$\frac{1}{8}$

Table 4.3

Y	0	1	2	3
$P(Y)$	$\frac{1}{8}$	$\frac{4}{8}$	$\frac{2}{8}$	$\frac{1}{8}$

The distribution of Z can be illustrated in a similar manner:

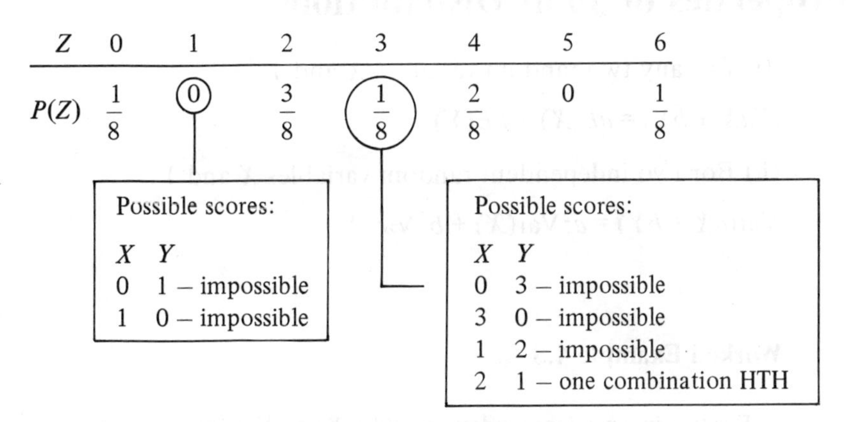

This information however is normally represented in the following tabular form (Table 4.4) and is called the joint probability distribution of X and Y. The column Sum is known as the *marginal distribution* of Y and the row Sum is the marginal distribution of X.

Table 4.4

$Y \backslash X$	*0*	*1*	*2*	*3*	*Sum*
0	1/8	0	0	0	1/8
1	0	3/8	1/8	0	4/8
2	0	0	2/8	0	2/8
3	0	0	0	1/8	1/8
Sum	1/8	3/8	3/8	1/8	

It is not difficult to see that the marginal distributions and the distributions of X and Y are one and the same.

Now $E(X) = 3/2$ and $E(Y) = 11/8$

and $E(Z) = \frac{3}{8} \times 2 + \frac{1}{8} \times 3 + \frac{2}{8} \times 4 + \frac{1}{8} \times 6 = \frac{23}{8} = E(X) + E(Y)$

$\text{Var}(X) = E(X^2) - 9/4 = 3/4$

$\text{Var}(Y) = E(Y^2) - 121/64 = 47/64$

and

$$\text{Var}(Z) = \frac{712}{64} - \frac{529}{64} = \frac{183}{64} \neq \text{Var}(X) + \text{Var}(Y)$$

Now two random variables X and Y are said to be independent if for all values x belonging to X and y belonging to Y,

$P(X = x \text{ and } Y = y) = P(X = x) \,.\, P(Y = y)$

Although intuitively X and Y are not independent in this problem, a counter-example would confirm that this is the case. This problem also illustrates that if two variables are not independent

$\text{Var}(Z) \neq \text{Var}(X) + \text{Var}(Y)$

4.4 Properties of Joint Distributions

(i) For any two random variables X and Y,

$E(aX + bY) = aE(X) + bE(Y)$

(ii) For two independent random variables X and Y,

$\text{Var}(aX + bY) = a^2\text{Var}(X) + b^2\text{Var}(Y)$

Worked Example 4.8

For two independent random variables X and Y show

$\text{Var}(X - Y) = \text{Var}(X) + \text{Var}(Y)$

Solution 4.8

$$\begin{aligned}\text{Var}(X - Y) &= \text{Var}(X) + \text{Var}(-Y)\\ &= \text{Var}(X) + \text{Var}(-1 \,.\, Y)\\ &= \text{Var}(X) + (-1)^2\,\text{Var}(Y)\\ &= \text{Var}(X) + \text{Var}(Y)\end{aligned}$$

The product of two random variables can be considered in a similar manner and the next example illustrates a further point.

Worked Example 4.9

Let X and Y be random variables defined as in Worked Example 4.7. Find $E(XY)$.

Solution 4.9

$$E(XY) = \frac{1}{8} \times 0 + \frac{3}{8} \times 1 + \frac{1}{8} \times 2 + \frac{2}{8} \times 4 + \frac{1}{8} \times 9$$

$$= 11/4 \neq E(X)E(Y)$$

It can be shown, however, that if X and Y are independent then

$E(XY) = E(X)E(Y)$

4.5 Covariance*

We have already seen that $\text{Var}(X + Y) \neq \text{Var}(X) + \text{Var}(Y)$ when X and Y are not independent. Investigating this further we will first of all make the algebra a little easier by using μ_X and μ_Y instead of $E(X)$ and $E(Y)$.

So $E(X + Y) = \mu_X + \mu_Y$

Now by definition

$$\begin{aligned}
\text{Var}(X + Y) &= E[((X + Y) - (\mu_X + \mu_Y))^2] \\
&= E[((X - \mu_X) + (Y - \mu_Y))^2] \\
&= E[(X - \mu_X)^2] + 2E[(X - \mu_X)(Y - \mu_Y)] + E[(Y - \mu_Y)^2] \\
&= \text{Var}(X) + 2E[(X - \mu_X)(Y - \mu_Y)] + \text{Var}(Y)
\end{aligned}$$

The expression $E[(X - \mu_X)(Y - \mu_Y)]$ is known as the *covariance* of X and Y and is usually written as $\text{Cov}(X, Y)$. Expanding this expression for $\text{Cov}(X, Y)$ we have

$$\begin{aligned}
&E[XY - X\mu_Y - \mu_X Y + \mu_X\mu_Y] \\
&\quad = E(XY) - \mu_Y E(X) - \mu_X E(Y) + \mu_X\mu_Y \\
&\quad = E(XY) - \mu_Y\mu_X - \mu_X\mu_Y + \mu_X\mu_Y \\
&\quad = E(XY) - \mu_X\mu_Y
\end{aligned}$$

From this last form of the covariance, it should be seen that if X and Y are independent then $\text{Cov}(X, Y) = 0$.

Worked Example 4.10

The random variable X has the following distribution:

X	0	1	2	3	4
$P(X)$	$c/4$	$c/3$	$c/3$	$c/2$	$c/12$

Find (i) c, (ii) $E(X)$, (iii) $\text{Var}(X)$

Solution 4.10

(i) $c\left(\frac{1}{4} + \frac{1}{3} + \frac{1}{3} + \frac{1}{2} + \frac{1}{12}\right) = 1$

$$\frac{18c}{12} = 1$$

$$\Rightarrow c = 2/3$$

(ii) $E(X) = \frac{1}{6} \times 0 + \frac{2}{9} \times 1 + \frac{2}{9} \times 2 + \frac{1}{3} \times 3 + \frac{1}{18} \times 4$

$= 17/9$

(iii) $\text{Var}(X) = E(X^2) - (17/9)^2$

$= \frac{1}{6} \times 0^2 + \frac{2}{9} \times 1^2 + \frac{2}{9} \times 2^2 + \frac{1}{3} \times 3^2 + \frac{1}{18} \times 4^2 - \frac{289}{81}$

$= 5 - 289/81$

$= 116/81$

Worked Example 4.11

A box contains 4 red balls and 2 blue balls. A ball is drawn without replacement until a blue ball is drawn. If X represents the number of draws required to draw a blue ball, find the probability distribution of X and hence calculate the mean and the variance.

Solution 4.11

X	1	2	3	4	5
Draw sequence:	B	RB	RRB	RRRB	RRRRB
Probabilities:	$\frac{1}{3}$	$\frac{2}{3}.\frac{2}{5}$	$\frac{2}{3}.\frac{3}{5}.\frac{2}{4}$	$\frac{2}{3}.\frac{3}{5}.\frac{2}{4}.\frac{2}{3}$	$\frac{2}{3}.\frac{3}{5}.\frac{2}{4}.\frac{1}{3}.1$
$P(X)$	$\frac{1}{3}$	$\frac{4}{15}$	$\frac{1}{5}$	$\frac{2}{15}$	$\frac{1}{15}$

$E(X) = \frac{1}{3} \times 1 + \frac{4}{15} \times 2 + \frac{1}{5} \times 3 + \frac{2}{15} \times 4 + \frac{1}{15} \times 5 = \frac{7}{3}$

$\text{Var}(X) = E(X^2) - 49/9$

$= \frac{1}{3} \times 1^2 + \frac{4}{15} \times 2^2 + \frac{1}{5} \times 3^2 + \frac{2}{15} \times 4^2 + \frac{1}{15} \times 5^2 - \frac{49}{9}$

$= 7 - 49/9$

$= 14/9$

Worked Example 4.12

The joint distribution of the random variables X and Y is given in Table 4.5.

Table 4.5

$Y \backslash X$	0	1	2
0	1/8	1/16	0
1	1/16	1/4	1/8
2	0	1/8	1/4

Find $E(X + Y)$.

Solution 4.12

We could tackle this problem in two ways:
Method (i): Directly from the table

$$E(X+Y) = \frac{1}{16} \times 1 + \frac{1}{16} \times 1 + \frac{1}{4} \times 2 + \frac{1}{8} \times 3 + \frac{1}{8} \times 3 + \frac{1}{4} \times 4$$

$$= 19/8$$

Method (ii): Finding the distributions of X and Y (marginal distributions)

X	0	1	2
$P(X)$	3/16	7/16	6/16

Y	0	1	2
$P(Y)$	3/16	7/16	6/16

So $E(X) = E(Y) = \frac{7}{16} \times 1 + \frac{6}{16} \times 2 = \frac{19}{16}$

$$E(X+Y) = E(X) + E(Y) = 19/8$$

Worked Example 4.13

Given a random variable X, which has a mean μ and variance σ^2, show that the random variable Z where $Z = (X - \mu)/\sigma$ has a mean of 0 and a variance of 1.

Solution 4.13

$$E(Z) = E\left(\frac{X-\mu}{\sigma}\right) = \frac{1}{\sigma}E(X-\mu) = \frac{1}{\sigma}[E(X) - \mu] = \frac{1}{\sigma}(\mu - \mu) = 0$$

$$\text{Var}(Z) = \text{Var}\left(\frac{X-\mu}{\sigma}\right) = \frac{1}{\sigma^2}\text{Var}(X) = \frac{\sigma^2}{\sigma^2} = 1$$

Worked Example 4.14

A random variable X has the probability distribution given in the table below, and $2Y = X - 18$. Find $E(Y)$ and $E(Y^2)$. Deduce the values of $E(X)$ and $\text{Var}(X)$.

X	12	16	18	20	24
$P(X)$	1/15	4/15	1/3	1/5	2/15

(L)

Solution 4.14

Y	-3	-1	0	1	3
$P(Y)$	1/15	4/15	1/3	1/5	2/15

$$E(Y) = \frac{1}{15} \times -3 + \frac{4}{15} \times -1 + \frac{1}{3} \times 0 + \frac{1}{5} \times 1 + \frac{2}{15} \times 3$$

$$= 2/15$$

$$E(Y^2) = \frac{1}{15} \times 9 + \frac{4}{15} \times 1 + \frac{1}{3} \times 0 + \frac{1}{5} \times 1 + \frac{2}{15} \times 9$$

$$= 34/15$$

Now as $2Y = X - 18$

$$E(2Y) = E(X - 18)$$

$$2E(Y) = E(X) - 18$$

So $E(X) = 18 + 4/15 = 18\frac{4}{15}$

and $\text{Var}(2Y) = \text{Var}(X - 18)$

$$4\text{Var}(Y) = \text{Var}(X)$$

Now $\text{Var}(Y) = E(Y^2) - [E(Y)]^2$

$$= \frac{34}{15} - \frac{4}{225} = \frac{506}{225}$$

$$\text{Var}(X) = 2024/225 = 9.00 \text{ (3 sig. figs)}$$

Worked Example 4.15

The discrete random variables X and Y, are independent and have the following probability distributions:

r	1	2	3	4	5
$P(X=r)$	0	1/4	1/3	1/3	1/12
$P(Y=r)$	1/5	0	2/5	0	2/5

Calculate the mean and variance of each of X and Y.
Give the probability distribution of $Z = X - Y$.
What are the mean and variance of Z? (OLE)

Solution 4.15

$$E(X) = 0 \times 1 + \frac{1}{4} \times 2 + \frac{1}{3} \times 3 + \frac{1}{3} \times 4 + \frac{1}{12} \times 5$$

$$= 13/4$$

$$E(Y) = \frac{1}{5} \times 1 + 0 \times 2 + \frac{2}{5} \times 3 + 0 \times 4 + \frac{2}{5} \times 5$$

$$= 17/5$$

$Z = X - Y$	−4	−3	−2	−1	0	1	2	3	4
$P(Z)$	0	1/10	2/15	7/30	1/6	11/60	1/10	1/15	1/60

Notation:
$P(X = i, Y = j) = p_{ij}$

$$\left.\begin{array}{l} X = 1, Y = 2, P_{12} = 0 \\ X = 2, Y = 3, P_{23} = 1/10 \\ X = 3, Y = 4, P_{34} = 0 \\ X = 4, Y = 5, P_{45} = 2/15 \end{array}\right\} = \frac{7}{30}$$

$$E(Z) = E(X - Y) = E(X) - E(Y) = -3/20$$

and because X and Y are independent,

$$\text{Var}(Z) = \text{Var}(X) + \text{Var}(Y)$$

$$\text{Var}(X) = 0 \times 1^2 + \frac{1}{4} \times 2^2 + \frac{1}{3} \times 3^2 + \frac{1}{3} \times 4^2 + \frac{1}{12} \times 5^2 - \frac{169}{16}$$

$$= 41/48$$

$$\text{Var}(Y) = \frac{1}{5} \times 1^2 + 0 \times 2^2 + \frac{2}{5} \times 3^2 + 0 \times 4^2 + \frac{2}{5} \times 5^2 - \frac{289}{25}$$

$$= 56/25$$

So

$$\text{Var}(Z) = 41/48 + 56/25 = 3713/1200 \quad (= 3.09)$$

Worked Example 4.16

Table 4.6 gives the joint probability distribution of two random variables X and Y; entries in the table represent $P(X = x \text{ and } Y = y)$

Table 4.6

$Y \backslash X$	-1	0	1
1	4/24	1/24	0
2	3/24	3/24	1/24
3	1/24	3/24	2/24
4	0	2/24	A

(a) What is the value of A?
(b) What is the probability that $X = 0$?
(c) Find $E(X)$ and $E(Y)$
(d) Find $E(XY)$
(e) Find $E(X + Y)$
(f) What is the conditional distribution of X given $Y = 1$?

(Give answers as fractions, not decimals.) (OLE)

Solution 4.16

(a) Sum of the probabilities in the table = 1

$$\Rightarrow A = 1 - 20/24 = 1/6$$

(b) From the marginal distribution of X,

$$P(X = 0) = \frac{1}{24} + \frac{3}{24} + \frac{3}{24} + \frac{2}{24} = \frac{3}{8}$$

(c) Marginal distributions of X and Y

X	-1	0	1
$P(X)$	8/24	9/24	7/24

Y	1	2	3	4
$P(Y)$	5/24	7/24	6/24	6/24

$$E(X) = \frac{-8}{24} + \frac{7}{24} = \frac{-1}{24}$$

$$E(Y) = \frac{5}{24} + \frac{14}{24} + \frac{18}{24} + \frac{24}{24} = \frac{61}{24}$$

(d) $E(XY) = \frac{1}{24} \times -3 + \frac{3}{24} \times -2 + \frac{4}{24} \times -1 + \frac{9}{24} \times 0 + \frac{1}{24} \times 2 + \frac{2}{24} \times 3 + \frac{4}{24} \times 4$

$= 11/24$

(e) $E(X+Y) = E(X) + E(Y) = (-1+61)/24 = 5/2$

(f)

X	-1	0	1
$P(X=x, Y=1)$	4/24	1/24	0
$P(Y=1)$	5/24	5/24	5/24
$P(X/Y=1)$	4/5	1/5	0

Worked Example 4.17*

Table 4.7 gives the joint probability distribution of two random variables X and Y. Show that X and Y are not independent but that $\text{Cov}(X, Y) = 0$.

Table 4.7

$Y \backslash X$	-1	0	1
0	0	1/8	0
1	1/4	0	1/4
2	0	3/8	0

*Solution 4.17**

Marginal distributions of X and Y

X	-1	0	1
$P(X)$	1/4	1/2	1/4

$E(X) = -1/4 + 1/4 = 0$

Y	0	1	2
$P(Y)$	1/8	1/2	3/8

$E(Y) = 1/2 + 3/4 = 5/4$

To show that two random variables are not independent we only require one example such that $P(X=x, Y=y) \neq P(X=x).P(Y=y)$

$P(X=-1, Y=0) = 0$

$P(X=-1).P(Y=0) = \frac{1}{4} \cdot \frac{1}{8} \neq 0$

$\Rightarrow X$ and Y are not independent

Now $\text{Cov}(X, Y) = E(XY) - E(X)E(Y)$

$= \frac{1}{8} \times 0 + \frac{1}{4} \times -1 + \frac{1}{4} \times 1 + \frac{3}{8} \times 0 - 0 \times \frac{5}{4}$

$= 0 - 0 = 0$

Worked Example 4.18*

n people are asked a question successively in a random order, and exactly 2 of the n know the answer. If $n > 5$, what is the probability that the first 4 of those asked do not know the answer?

Show that the probability that the rth person asked is the first to know the answer is

$$\frac{2(n-r)}{n(n-1)} \text{ if } 1<r<n$$

Show that this expression for the probability also holds when $r = 1$ or $r = n$.

Verify that the sum of these probabilities over all possible values of r is 1, and show that the expected number of people to be asked before the correct answer is obtained is $(n + 1)/3$.

(OLE)

*Solution 4.18**

P(first 4 of those asked do not know the answer)

$$= \frac{(n-2)}{n} \cdot \frac{(n-3)}{(n-1)} \cdot \frac{(n-4)}{(n-2)} \cdot \frac{(n-5)}{(n-3)}$$

P(rth person asked is the first to know the answer)

= P(first $(r-1)$th persons asked do not know the answer and rth person knows the answer)

$$= \frac{(n-2)}{n} \cdot \frac{(n-3)}{(n-1)} \cdots\cdots \frac{(n-r)}{(n-r+2)} \cdot \frac{2}{(n-r+1)}$$

$$= \frac{2(n-r)}{n(n-1)}$$

When $r = 1$, probability $= \dfrac{2(n-1)}{n(n-1)} = \dfrac{2}{n}$

which is simply the probability that the first person chosen knows the answer.

When $r = n$, probability $= \dfrac{2(n-n)}{n(n-1)} = 0$

As the nth person is the last person, one who knows the answer will already have answered the question, so the nth person cannot possibly be the first person to answer the question correctly.

$$\sum_{r=1}^{r=n-1} \frac{2(n-r)}{n(n-1)} = \frac{2}{n(n-1)} \sum_{r=1}^{r=n-1} (n-r)$$

$$= \frac{2}{n(n-1)} \left\{ \sum_{r=1}^{r=n-1} n - \sum_{r=1}^{r=n-1} r \right\}$$

$$= \frac{2}{n(n-1)} [n(n-1) - n(n-1)/2]$$

$$= \frac{2}{n(n-1)} \frac{(n-1)}{2} [2n - n]$$

$$= 1$$

$$E(X) = \sum_{r=1}^{r=n-1} \frac{2r(n-r)}{n(n-1)}$$

$$= \frac{2}{n(n-1)} \sum_{r=1}^{r=n-1} (rn - r^2)$$

$$= \frac{2}{n(n-1)} [\tfrac{1}{2}nn(n-1) - \tfrac{1}{6}(n-1)n(2n-1)]$$

$$= \frac{2n(n-1)}{n(n-1)} . \tfrac{1}{6}[3n - (2n-1)]$$

$$= (n+1)/3$$

4.6 Exercises

Exercise 4.1

Two fair dice are thrown and the score X is found by multiplying the two scores shown on the dice together. Draw up a table showing the probability distribution of X.

Find (i) $P(X > 12)$
(ii) $E(X)$
(iii) $\text{Var}(X)$
(iv) Given that $X > 12$, find $P(X > 18)$.

Exercise 4.2

A discrete random variable X takes integer values between 0 and 5 inclusive with probabilities given by

$$P(X = r) = \begin{cases} \dfrac{2r+1}{20} & r = 0, 1, 2, 3 \\[2ex] \dfrac{11-2r}{20} & r = 4, 5. \end{cases}$$

Find the expectation and variance of X.

Exercise 4.3

There are 4 coins in a bag. A £1, 2 × 50p and a 20p coin. Find the expected amount drawn if (i) one coin is drawn, (ii) two coins are drawn without replacement.

Exercise 4.4

Three fair coins are thrown. A score of one is recorded for each head and a score of two for each tail. If the three scores are multiplied together, find the mean and variance of the total score.

Exercise 4.5

The discrete random variable X has the distribution

X	1	2	3	4
$P(X)$	1/10	2/5	1/4	1/4

Find the mean and variance of the distributions (i) $X + 2$, (ii) $3X - 2$.

Exercise 4.6

If X and Y are independent discrete random variables, show that

$$E(XY) = E(X)E(Y)$$

Exercise 4.7

The joint probability distribution of the discrete random variables X and Y are shown in Table 4.8.

Table 4.8

$Y \backslash X$	−2	0	2
−1	1/8	1/4	1/8
1	1/16	3/8	1/16

Show that X and Y are not independent and verify that $\text{Cov}(X, Y) = 0$.

Exercise 4.8

Using the data of Worked Example 4.7, verify that

$$\text{Var}(X + Y) = \text{Var}(X) + 2\text{Cov}(X, Y) + \text{Var}(Y)$$

Exercise 4.9

Two fair dice are thrown. If X is the score of the first die and Y is the smaller of the two scores, tabulate the joint distribution of X and Y. Find (i) $E(Y)$, (ii) $\text{Var}(Y)$, (iii) $E(XY)$, (iv) $\text{Cov}(X, Y)$.

Exercise 4.10

A cubical die has two each of its faces numbered 1, 2, and 3, respectively, and is such that the probabilities of obtaining these scores in a single throw are 0.1, 0.8 and 0.1, respectively.

(i) If X is the score obtained in one throw of the die, determine the mean and the variance of X.

(ii) Let M denote the median of the three scores obtained in three independent throws of the die. Show that $P(M = 1) = 0.028$. Evaluate $P(M = 2)$ and $P(M = 3)$, and hence determine the mean and the variance of the sampling distribution of M. (WJEC)

Exercise 4.11*

An unbiased coin is tossed four times. Let the random variable X denote the greatest number of successive heads occurring in the four tosses (e.g. if HTHH occurs, then $X = 2$, but if TTHT occurs, then $X = 1$). Show that $E(X) = 27/16$ and find $\text{Var}(X)$.

The random variable Y is the number of heads occurring in the four tosses. Find $\text{Cov}(X, Y)$. (UCLES)

Exercise 4.12

A cubical die has three faces marked with a '1', two faces marked with a '2' and one face marked with a '3'. Calculate the expectation and variance of the score obtained when this die is thrown once.

Deduce, or find otherwise, the expectation and variance of the score obtained in one throw of a second cubical die, which has one face marked '1', two faces marked '2' and three faces marked '3'.

Two of the first type of die and one of the second type are thrown together, and X denotes the total score obtained. Denoting the expectation and variance of X by μ and σ^2 respectively, show that

(i) $\sigma^2 = 5/3$, (ii) $P(|X - \mu| > 2\sigma) = 1/18$. (UCLES)

Exercise 4.13

In the village of Jieyuhao families are strictly limited to 2 children. The probability distribution of the number of children of any given individual is given in Table 4.9.

Table 4.9

Number of children	0	1	2
Probability	1/6	1/2	1/3

Find the mean and variance of the number of children.

Assuming that individuals are independent, find the probability distribution of the number of grandchildren of a given individual. What is the expected number of grandchildren? Comment on the assumption of independence. (OLE)

Exercise 4.14*

A discrete random variable X takes values $-1, 0, 1$, and a discrete random variable Y takes values $0, 1, 2$. The joint probability distribution of X and Y is given in Table 4.10 where $p \neq 0$.

Table 4.10

X \ Y	0	1	2
−1	p	0	p
0	q	p	q
1	p	0	p

(i) State, giving a reason, whether X and Y are independent.
(ii) Show that $\text{Cov}(X, Y) = 0$.
(iii) Given that $\text{Var}(Y) = 5/6$, find the values of p and q.
(iv) With p, q having the values found in (iii), find the conditional distribution of X for each value of Y. (UCLES)

Exercise 4.15*

An unbiased cubical die is thrown twice. Let X_1 be the score on the first throw and X_2 the score on the second throw. Random variables Z and W are defined by

$$Z = X_1 + X_2$$

$$W = 3X_1 - 2X_2$$

Find (i) $E(W)$.
(ii) $\text{Var}(W)$.
(iii) $\text{Cov}(X_1, X_2)$, where $\text{Cov}(X_1, X_2)$ is the covariance of X_1 and X_2.
(iv) $\text{Cov}(X_1, Z)$. (UCLES)

Exercise 4.16*

When a market gardener takes n cuttings from a shrub and plants them, the number X that will root successfully is a discrete random variable with

$$P(X = r) = \frac{2r}{n(n+1)}, r = 1, 2, \ldots, n$$

The total cost in pence to the gardener of taking and planting n cuttings is equal to $20 + 0.8n^2$. Cuttings that root successfully are sold by the gardener for 60 pence each.

(a) Show that $E(X) = (2n + 1)/3$
(b) For $n = 20$, calculate
  (i) the probability that the gardener will make a loss,
  (ii) the gardener's expected profit.
(c) Determine the value of n which will maximise the gardener's expected profit and evaluate this maximum expected profit. (WJEC)

5 Binomial and Poisson Distributions

5.1 Definition of the Binomial Distribution

If the probability that an event or trial happens is p, then the probability that it does not happen is $1 - p$, denoted by q. If the event is repeated n times, the probability that the event occurs exactly r times, i.e. $P(X = r)$ is given by

$$P(X = r) = \binom{n}{r} p^r q^{n-r} \qquad r = 0, 1, 2, \ldots$$

Worked Example 5.1

A fair coin is tossed 5 times, find:

(i) the probability of obtaining 3 heads: (ii) the probability of obtaining at least 3 heads.

Solution 5.1

(i) This is a straightforward use of the above formula.

$$P(X = 3) = \binom{5}{3} (\tfrac{1}{2})^3 (\tfrac{1}{2})^2 \quad \text{since } p = \tfrac{1}{2}$$

$$= \frac{5!}{3!2!} \times \frac{1}{32} = \frac{5}{16}$$

(ii) The event of obtaining at least 3 heads is made up of the events of obtaining 3, 4 or 5 heads.

Hence, $P(X \geqslant 3) = P(X = 3) + P(X = 4) + P(X = 5)$

$$= \binom{5}{3} (\tfrac{1}{2})^3 (\tfrac{1}{2})^2 + \binom{5}{4} (\tfrac{1}{2})^4 (\tfrac{1}{2}) + (\tfrac{1}{2})^5$$

$$= \frac{10}{32} + \frac{5}{32} + \frac{1}{32} = \frac{1}{2}$$

The formula is a statement of the *binomial probability distribution.*

5.2 Shape of the Distribution

The binomial distribution is an example of a discrete distribution, because the variable x can only take integer values. Figure 5.1 shows 3 typical distributions for $n = 10$, and $p = \frac{1}{4}, \frac{1}{2}, \frac{3}{4}$ respectively.

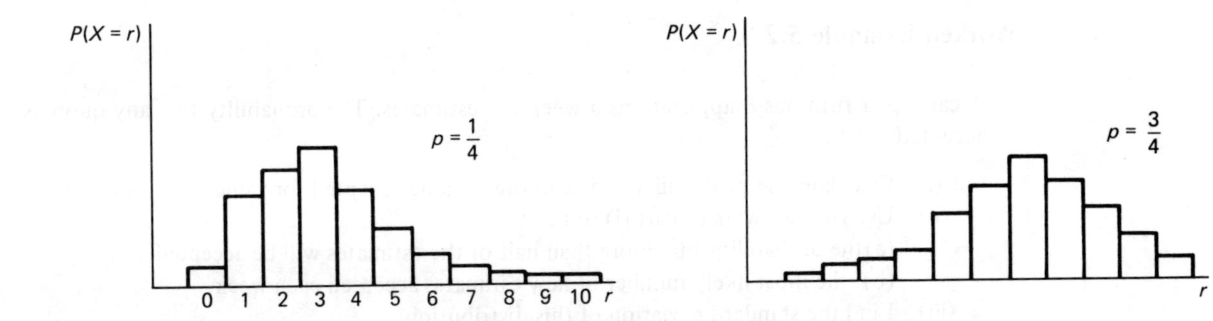

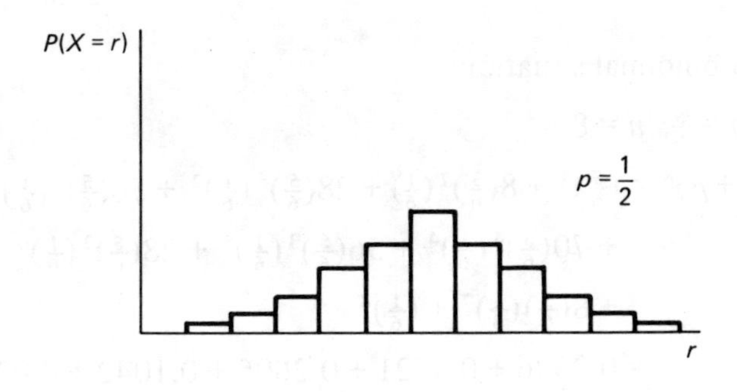

Figure 5.1

We can see that $p < \frac{1}{2}$ skewed to the right;

$p = \frac{1}{2}$ symmetrical;

$p > \frac{1}{2}$ skewed to the left.

5.3 Mean and Variance

If X is a discrete random variable with a binomial probability distribution, then the mean of X or expectation of X, $E(X)$ is given by

$$E(X) = \sum_{r=0}^{n} r p_r$$

$$= \sum_{r=0}^{n} r \binom{n}{r} p^r q^{n-r}$$

It can be shown that this sum is equal to np (see Worked Example 8.10). Hence, $E(X) = np = \mu$.

The variance is given by

$$\text{Var}(X) = E(X^2) - \{E(X)\}^2$$

$$= \sum_{r=0}^{n} r^2 \binom{n}{r} p^r q^{n-r} \quad - n^2p^2$$

$$= \cancel{n^2p^2} - np^2 + np - \cancel{n^2p^2} \quad \text{(see Worked Example 8.10)}$$

$$= np(1-p) = npq$$

Hence, $\text{Var}(X) = npq$

Worked Example 5.2

A car repair firm has 8 applications a week for estimates. The probability that any quote is accepted is 1/6.

(i) Calculate the probability that x quotes will be accepted for values of x from 0 to 8.
(ii) Use your answers to Part (i) to find:
(a) the probability that more than half of the estimates will be accepted;
(b) the most likely number of new estimates accepted each week.
(iii) Find the standard deviation of this distribution.

Solution 5.2

This is a binomial situation.

$p = \frac{1}{6}$, $q = \frac{5}{6}$, $n = 8$

(i) $(q + p)^8 = (\frac{5}{6})^8 + 8(\frac{5}{6})^7(\frac{1}{6}) + 28(\frac{5}{6})^6(\frac{1}{6})^2 + 56(\frac{5}{6})^5(\frac{1}{6})^3$
$+ 70(\frac{5}{6})^4(\frac{1}{6})^4 + 56(\frac{5}{6})^3(\frac{1}{6})^5 + 28(\frac{5}{6})^2(\frac{1}{6})^6$
$+ 8(\frac{5}{6})(\frac{1}{6})^7 + (\frac{1}{6})^8$
$= 0.2326 + 0.3721 + 0.2605 + 0.1042 + 0.026$
$+ 0.0042 + 0.0004 + 0.00002 + 0.0000006$

$\Rightarrow$ Probabilities are given by

x	0	1	2	3	4	5	6	7	8
P	0.2326	0.3721	0.2605	0.1042	0.026	0.0042	0.0004	0.00002	0.0000006

(ii) (a) P(more than 4 accepted) $= 0.0042 + \ldots + 0.0000006$
$= 0.0046$

(b) the highest probability is 0.3721
$\Rightarrow$ The most likely number accepted is 1.

(iii) Standard deviation $= \sqrt{npq} = \sqrt{8 \times \frac{5}{6} \times \frac{1}{6}} = 1.05$

Worked Example 5.3

A random variable X is binomially distributed with mean 12.38 and variance 8.64. Find n and p.

Solution 5.3

Using $\mu = np$ and $\sigma^2 = npq$, we have

$12.38 = np$ (1)
$8.64 = npq$ (2)

replacing np in (2), gives $8.64 = 12.38q$

$\Rightarrow q = \dfrac{8.64}{12.38} = 0.698$

$\Rightarrow p = 1 - q = 0.302$

from (1), $n = \dfrac{12.38}{0.302} = 41$

5.4 Expected Frequency Distribution

Sometimes it is required to fit a theoretical distribution to a set of data obtained by experiment. The following worked example illustrates how to fit a binomial distribution.

Worked Example 5.4

A random variable X which can take the value 0, 1, 2, 3, 4, was found to have the distribution shown in Table 5.1. Fit a binomial distribution to this data.

Table 5.1

X	0	1	2	3	4
f	28	70	50	8	4

Solution 5.4

The mean value of X is:

$$\frac{\Sigma fx}{\Sigma f} = \frac{0 \times 28 + 1 \times 70 + 2 \times 50 + 3 \times 8 + 4 \times 4}{160}$$

$$= 1.31$$

Since the mean of a binomial distribution is np, we have:

$$np = 1.31$$

$$\Rightarrow p = \frac{1.31}{4} = 0.328 \Rightarrow q = 0.672$$

$$\begin{aligned}(q+p)^4 &= (0.672)^4 + 4(0.672)^3(0.328) + 6(0.672)^2(0.328)^2 \\ &\quad + 4(0.672)(0.328)^3 + (0.328)^4 \\ &= 0.204 + 0.398 + 0.291 + 0.095 + 0.012\end{aligned}$$

Therefore probabilities are

x	0	1	2	3	4
P	0.204	0.398	0.291	0.095	0.012

The table gave values for 160 trials of the experiment, hence to calculate the frequencies, the probabilities must be multiplied by 160.

The frequencies are 32, 64, 47, 15, 2. How good these theoretical frequencies are in predicting the observed results, can be tested using the chi squared test given in Chapter 12.

Worked Example 5.5

An arcade game is such that the probability of any person winning is always 0.3. How many people need to play the game, to ensure that the probability that at least one person wins is 0.96.

Solution 5.5

This is a binomial situation, $p = 0.3$, $q = 0.7$, but n is unknown.

Now $P(X \geqslant 1)$ can be written as $1 - P(X = 0) = 1 - (0.3)^n$.

$$\Rightarrow 1 - (0.3)^n = 0.96$$

$$\Rightarrow (0.3)^n = 0.04$$

Take logs $n \log 0.3 = \log 0.04$

$$\Rightarrow n = \frac{\log 0.04}{\log 0.3} = 2.7$$

Hence 3 people need to play the game.

5.5 The Poisson Distribution

The discrete probability distribution defined by

$P(X = r) = \dfrac{\lambda^r e^{-\lambda}}{r!}$, $r = 0, 1, 2, \ldots$ where λ is a given constant called the Poisson distribution.

The mean of the distribution is given by

$$\text{mean} \sum_{r=0}^{\infty} \frac{r\lambda^r e^{-\lambda}}{r!} = \lambda e^{-\lambda} \sum_{r=1}^{\infty} \frac{\lambda^{r-1}}{(r-1)!}$$

$$= \lambda e^{-\lambda} \left[1 + \lambda + \frac{\lambda^2}{2!} + \ldots\right] = \lambda e^{-\lambda} . e^{\lambda} = \lambda.$$

Hence λ is the mean of the distribution.

It can be shown that the variance σ^2 is given by

$$\sigma^2 = \sum_{r=0}^{\infty} \frac{r^2 \lambda^r e^{-\lambda}}{r!} - \lambda^2$$

$= \lambda$ (see Worked Example 8.12)

The main use of this distribution is in problems dealing with rare isolated events which occur independently and randomly in continuous situations such as space, length or time.

5.6 Poisson Distribution as an Approximation to the Binomial Distribution

If X is a random variable and the probability of success p in n trials of X is very small, then the binomial theorem becomes difficult to handle if n is large. Consider the following example.

Worked Example 5.6

A book contains 80 pages. Each page on average contains 300 words, and the total number of typesetting errors in the book is 200. What is the probability that on any page, 4 or more errors have to be corrected?

Solution 5.6

If we use the binomial distribution, then $n = 300$, $p = \dfrac{200}{80 \times 300}$

$$= 0.0083$$

$\Rightarrow q = 0.9917$

$$p(X \geqslant 4) = 1 - \left[\binom{200}{0} (0.9917)^{200} + \binom{200}{1} (0.0083)\ (0.9917)^{199}\right.$$

$$\left. + \ldots + \binom{200}{3} (0.0083)^3 (0.9917)^{197}\right]$$

To evaluate this expression is clearly very difficult. It can be shown, however, that if n is large ($n \geqslant 50$), p very small, and $np = k$ where $k \leqslant 5$, then the binomial distribution can be approximated by the Poisson distribution, where

$$P(X = r) = \frac{k^r e^{-k}}{r!}$$

Worked Example 5.7

Repeat Worked Example 5.6 using the Poisson distribution.

Solution 5.7

$k = 300 \times 0.008\dot{3} = 2.5$

$$\Rightarrow P(X \geqslant 4) = 1 - \left[e^{-2.5} + 2.5\, e^{-2.5} + \frac{2.5^2 e^{-2.5}}{2!} + \frac{2.5^3 e^{-2.5}}{3!}\right]$$

$$= 1 - e^{-2.5} \left[1 + 2.5 + \frac{2.5^2}{2} + \frac{2.5^3}{6}\right]$$

$$= 0.242$$

5.7 Additive Property of the Poisson Distribution

If X_1 is a random variable following a Poisson distribution with mean λ_1, and X_2 is a random variable independent of X_1, following a Poisson distribution with mean λ_2, then $X_1 + X_2$ also follows a Poisson distribution with mean $\lambda_1 + \lambda_2$.

Worked Example 5.8

A warehouse supplies 2 stores with crates of HOLDFAST glue. The number of crates per week needed by the two stores follow Poisson distributions with means 0.6 and 0.1 respectively.

(i) Calculate the probability that two or more crates in total are needed from the warehouse in any week.

(ii) What is the minimum stock requirement needed in the warehouse for which there is a probability greater than 0.9 that all the stores' requirements can be met.

Solution 5.8

(i) The requests for each store are presumed independent, hence the total follows a Poisson distribution with mean 0.6 + 0.1 = 0.7
$P(2 \text{ or more}) = 1 - P(0) - P(1)$
$= 1 - e^{-0.7} - 0.7e^{-0.7} = 0.156$

(ii) The probability distribution is

$$P(X) = e^{-0.7} + 0.7e^{-0.7} + \frac{0.7^2 e^{-0.7}}{2!} + \ldots$$

i.e.

	0	1	2
$P(X)$	0.497	0.348	0.122

The sum of the first three terms is greater than 0.9. Hence two crates are required to supply demands.

Worked Example 5.9

(a) Berg and Korner have a long rivalry in tennis: in the last 25 games, Berg has won 15 times. They start a new series of 8 games. Assuming that the binomial model may be applied, what is the probability that Berg will win at least 6 of these?

(b) Prove that the mean of the binomial distribution is np. A set of 100 pods, each containing 4 peas, was examined to see how many of the peas were good. The following were the results:

No. of good peas in pod	0	1	2	3	4
No. of pods	7	20	35	30	8

(i) What was the probability of getting a good pea?
(ii) Calculate the theoretical frequencies of 0, 1, 2, 3, 4 good peas, using the associated theoretical binomial distribution. (SUJB)

Solution 5.9

(a) The probability that Berg wins a game $P = \frac{15}{25} = 0.6$

$$\Rightarrow P(\text{at least six wins}) = P(6 \text{ wins}) + P(7 \text{ wins}) + P(8 \text{ wins})$$
$$= {}^8C_6\, 0.6^6 \times 0.4^2 + {}^8C_7\, 0.6^7 \times 0.4 + 0.6^8$$
$$= 0.2090 + 0.0896 + 0.0170$$
$$= 0.3156$$

(b) The first part of this question has already been proved.

(i) The number of good peas found was

$$0 \times 7 + 1 \times 20 + 2 \times 35 + 3 \times 30 + 4 \times 8 = 212$$

The number of peas examined = 100 × 4 = 400

$$\Rightarrow \text{probability that a pea is good} = \frac{212}{400} = 0.53$$

The relevant binomial distribution is $(0.47 + 0.53)^4$

$= 0.47^4 + 4 \times 0.47^3 \times 0.53 + 6 \times 0.47^2 \times 0.53^2 + 4 \times 0.47 \times 0.53^3 + 0.53^4$

$= 0.049 + 0.220 + 0.372 + 0.280 + 0.079$

The terms in this expansion must be multiplied by 100 to get the frequencies, which to the nearest whole numbers are

5, 22, 37, 28, 8

Worked Example 5.10

Show that, for the Poisson distribution in which the probabilities of 0, 1, 2, . . . successes are e^{-m}, me^{-m}, $\dfrac{m^2e^{-m}}{2!}$, . . . , the mean number of successes is equal to m. State the variance.

A sales manager receives 6 telephone calls on average between 9.30 a.m. and 10.30 a.m. on a weekday. Find the probability that:

(i) he will receive 2 or more calls between 9.30 and 10.30 on a certain weekday;
(ii) he will receive exactly 2 calls between 9.30 and 9.40;
(iii) during a normal 5-day working week, there will be exactly 3 days on which he will receive no calls between 9.30 and 9.40. (SUJB)

Solution 5.10

The first part is covered in the text.

(i) Considering our unit of time to be 1 hour, the average number of calls per hour is 6. Hence $m = 6$.

$\Rightarrow P(2 \text{ or more}) = 1 - P(0) - P(1)$

$= 1 - e^{-6} - 6e^{-6} = 0.9826$

(ii) The time interval is now 10 mins. If we assume the calls are evenly spread throughout the hour, then the number of calls on average in 10 mins is 1

$\Rightarrow m = 1$

$\Rightarrow P(2) = \dfrac{1^2 \times e^{-1}}{2!} = 0.1839$

(iii) Consider the five days as being a binomial situation with $p = P$ (no calls between 9.30 and 9.40)

$= e^{-1} = 0.3679, \; q = 0.6321$

$\Rightarrow P(3 \text{ days}) = {}^5C_3\,(0.6321)^2(0.3679)^3 = 0.199$

5.8 Exercises

Exercise 5.1

In a large consignment of packets of sugar, 5% of the packets are underweight. Random samples, each consisting of 8 packets, are examined. Calculate the mean and the standard deviation of the number of underweight packets per sample.

If 120 such samples are examined, calculate the expected number of samples which contain not more than one underweight packet. (UCLES)

Exercise 5.2

A multiple choice test consists of 30 questions, each having 4 possible answers, only one of which is correct. A group of students take the test. On the assumption that they guess the answer:

(i) Calculate their mean score.
(ii) Calculate the standard deviation of their scores.
(iii) Find the greatest score that one of these candidates might be expected to obtain.

Exercise 5.3

In an experiment a certain number of dice are thrown and the number of sixes obtained is recorded. The dice are all biased and the probability of obtaining a six with each individual die is p. In all there were 60 experiments and the results are shown in the table.

Number of sixes obtained in an experiment	0	1	2	3	4	More than 4
Frequency	19	26	12	2	1	0

Calculate the mean and the standard deviation of these data. By comparing these answers with those expected for a binomial distribution, estimate:

(i) the number of dice thrown in each experiment;
(ii) the value of p.

Exercise 5.4

An examination consists of eight questions in each of which the candidate must say which one of five answers is the correct one. On the assumption that a candidate who has done no preparatory work chooses for each question any one of the five answers with equal probability,

(a) prove that the probability that he gets more than one correct answer is equal to $\dfrac{5^8 - 3 \times 4^8}{5^8}$
(b) find the probability that he gets correct answers to six or more questions. (L)

Exercise 5.5

During an influenza epidemic, 6% of the population of a large city were affected on a given day. In a factory employing 60 people, calculate:

(i) the probability that on a given day, no more than one employee will be absent with 'flu';
(ii) the probability that on a given day exactly four employees will be absent with 'flu'.

Use the Poisson approximation to the binomial distribution, to calculate the probability that at least 10 employees will be absent.

Exercise 5.6

Derive the mean and variance of the binomial distribution.

(a) A girls' netball team has a constant probability of 0.8 of winning a home match and 0.6 of winning an away match. During the season, the team plays six home games and six away games.

Calculate the mean and the variance of the number of matches that will be won by the team during the season.

(b) Packets of Chewy Fruits contain eight sweets, and two flavours (raspberry and blackcurrant) are made. The filling machine mixes the two types in the ratio 2 : 1, respectively. What is the probability that any packet contains at least one raspberry and at least three blackcurrant sweets?

Exercise 5.7

A firm specialises in hiring out microcomputers. It has 10 machines available and hiring contracts are made on a day-to-day basis. It is found by experience that the demand follows a Poisson distribution with mean 6.5.

(i) What is the probability that no machines will be on loan on a given day?
(ii) Calculate the proportion of days on which a request for hire will be refused.
(iii) What is the most likely number of machines to be in use on a given day.
(iv) Calculate the expected income per year (300 days) if the hiring charge per day is £8.
(v) How many machines are necessary if the probability of not meeting demand is to be no more than 0.05.

Exercise 5.8

A random variable X has a Poisson distribution given by

$Pr(X = r) = p_r = e^{-\lambda} \lambda^r / r!, r = 0, 1, 2, \ldots$

Prove that the mean of X is λ. Give two examples (other than that suggested below) of situations where you would expect a Poisson distribution to occur.

The number of white corpuscles on a slide has a Poisson distribution with mean 3.2. By considering the values of r for which $p_{r+1}/p_r > 1$ find the most likely number of white corpuscles on a slide. Calculate correct to 3 decimal places the probability of obtaining this number. If two such slides are prepared what is the probability, correct to three decimal places, of obtaining at least two white corpuscles in total on the two slides.

(SUJB)

Exercise 5.9

(a) The random variable X has a Poisson distribution and is such that $P(X = 2) = 3P(X = 4)$. Find, correct to three decimal places, the values of (i) $P(X = 0)$, (ii) $P(X \leqslant 4)$.
(b) The number of characters that are mistyped by a copytypist in any assignment has a Poisson distribution, the average number of mistyped characters per page being 0.8. In an assignment of 80 pages calculate, to three decimal places,

(i) the probability that the first page will contain exactly two mistyped characters;

(ii) the probability that the first mistyped character will appear on the third page. (WJEC)

Exercise 5.10

Two independent random variables X_1 and X_2 have Poisson distributions with means λ_1 and λ_2 respectively. What is the mean and variance of Z where $Z = X_1 + X_2$?

The probability distributions of the number of daily absences of male and female employees in a factory can be approximated by a Poisson model with mean 0.8 for males and 0.6 for females. Assuming that the sum of two independent Poisson variables is a Poisson variable, calculate, correct to three decimal places, the probability that

(i) two men and two women are absent on the same day;

(ii) four employees are absent on any one day;

(iii) during a working week of five days there are no more than two employees absent. (SUJB)

Exercise 5.11

(a) Two independent random variables, X and Y, have Poisson distributions with means μ and 2μ, respectively. Show that for every non-negative integer r,

$$P(X = r, Y = r + 2) = 4P(X = r + 2, Y = r)$$

(b) During its first year in operation the number of occasions that a television set will need to be serviced has a Poisson distribution with mean 2.4. Independently of what happened in the first year, the number of occasions the set will need to be serviced during its second year in operation has a Poisson distribution with mean 4.6. During its third year in operation the number of occasions the set will need to be serviced is exactly twice the number of occasions it was serviced in its second year.

(i) Without assuming any result regarding Poisson distributions that is not given in the information booklet, find the probability, to three decimal places, that the set will need to be serviced a total of 2 or more occasions during its first two years in operation.

(ii) Find, correct to three decimal places, the probability that the set will need to be serviced a total of 4 or more occasions during its second and third years in operation.

Determine the mean and the variance of the total number of occasions that the set will need to be serviced

(iii) during its first two years in operation;

(iv) during its first three years in operation. (WJEC)

Exercise 5.12

A computer has been programmed to select, at random, either the value $\mu = 1$ or the value $\mu = 2$ (both values being equally likely), and then to generate a sequence of values from a Poisson distribution whose mean is the value of μ selected. You may observe as many values in the sequence as you like, but you must pay £1 for each value observed; at any time, you may stop observing, and guess at the value of μ. If you are correct, you are given £5, and after your guess the computer selects, at random, a fresh value of μ for the next game. Your policy is to select a value of $n (n \geqslant 1)$, observe the first n values in the sequence, and guess '$\mu = 1$' if the mean of these values is strictly less than 1.5, otherwise you guess '$\mu = 2$'.

The digits below are a string of values from a Poisson distribution of mean 1, grouped in blocks of four for convenience:

3110 2011 2101 0231 0012 0102 3011 1002 1010 4011

(i) Explain how they can be used to obtain values from a Poisson distribution of mean 2.

(ii) Describe a method you can use, in an examination room, to simulate the computer's choice of μ.

(iii) Use (i) and (ii) to simulate 3 plays of the game for each of the values $n = 1, 2$ and 3 (i.e. 9 games in all), setting out your results clearly.

(iv) Describe how to use a more extensive simulation to select the best value of n for your policy. (OLE)

Exercise 5.13

In a certain country it is known that 35% of the adult population have some knowledge of a foreign language. If 10 adults from this country are chosen at random, find the probability that

(i) at least one of those chosen will have some knowledge of a foreign language,

(ii) at most three of those chosen will have some knowledge of a foreign language.

For one particular foreign language, only a very small proportion $r\%$ of the adult population have some knowledge of it. It is required to select n adults at random, where n is chosen so that the probability of obtaining at least one adult having some knowledge of the language is to be 0.99, as nearly as possible. Use a suitable Poisson approximation to show that $n \approx \dfrac{460.5}{r}$.

For the case when $r = \frac{1}{2}$ and $n = 921$, find the probability that precisely four adults in the sample will have some knowledge of the language. (UCLES)

Exercise 5.14

A hockey team consists of 11 players. It may be assumed that, on every occasion, the probability of any one of the regular members of the team being unavailable for selection is 0.15, independently of all other members. Calculate, giving three significant figures in your answers, the probability that, on a particular occasion,

(i) exactly one regular member is unavailable,

(ii) more than two regular members are unavailable.

Taking the probability that more than 3 regular members are unavailable as 0.07, write down, for a season in which 50 matches are played, the expected value of the number of matches for which more than 3 regular members are unavailable. Use a suitable Poisson distribution to find an approximation for the probability that, in the course of a season, more than 3 regular members will be unavailable at most twice. (UCLES)

6 Continuous Probability Distributions

6.1 Continuous Random Variables

In Chapters 4 and 5 we have dealt with a finite or countably infinite number of outcomes to a random experiment. When however we have an outcome space that cannot be counted, a random variable X is a measure on this outcome space. In this instance X is said to be a *continuous random variable* and rather than summing over a finite number of values to determine parameters such as the mean or variance, the calculus of integration is used instead.

If X is a continuous random variable and $f(x)$ is a function defined on X such that

$$P(a < X < b) = \int_a^b f(x)\mathrm{d}x$$

and the following conditions are satisfied:

(i) $f(x) \geqslant 0$ for all possible values of x

(ii) $\displaystyle\int_{-\infty}^{\infty} f(x)\mathrm{d}x = 1$

then $f(x)$ is said to be a *probability density function* of X.

Clearly $P(X = a) = \displaystyle\int_a^a f(x)\mathrm{d}x = 0$

– probabilities are assigned to intervals, not to specific values.

Worked Example 6.1

The continuous random variable X has probability density function $(x - a)(2a - x)$ for $a \leqslant x \leqslant 2a$, and zero elsewhere. Sketch the graph of the function and show that $a^3 = 6$. (OLE modified)

Solution 6.1

Although the function is only greater than zero for values of x in the interval $a < x < b$, it is defined on the range $-\infty$ to $+\infty$ and this should be made clear on any diagram. In this example $f(x) = 0$ for values of x less than a and greater than $2a$. In the range $a \leqslant x \leqslant 2a$ the function is parabolic.

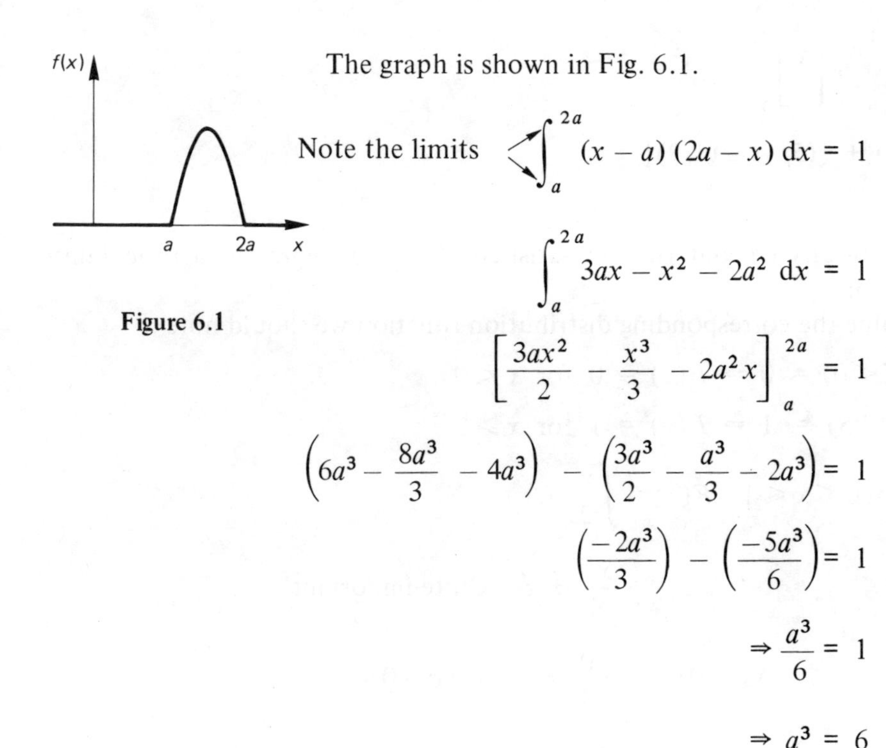

Figure 6.1

The graph is shown in Fig. 6.1.

Note the limits
$$\int_a^{2a} (x-a)(2a-x)\,\mathrm{d}x = 1$$

$$\int_a^{2a} 3ax - x^2 - 2a^2\,\mathrm{d}x = 1$$

$$\left[\frac{3ax^2}{2} - \frac{x^3}{3} - 2a^2x\right]_a^{2a} = 1$$

$$\left(6a^3 - \frac{8a^3}{3} - 4a^3\right) - \left(\frac{3a^3}{2} - \frac{a^3}{3} - 2a^3\right) = 1$$

$$\left(\frac{-2a^3}{3}\right) - \left(\frac{-5a^3}{6}\right) = 1$$

$$\Rightarrow \frac{a^3}{6} = 1$$

$$\Rightarrow a^3 = 6$$

6.2 Cumulative Distribution Function

The probability density function $f(x)$ is not in itself a probability, however we can represent probabilities in a function form by defining the *cumulative distribution function* $F(x)$

$$F(x) = P(X \leqslant x) = \int_{-\infty}^{x} f(x)\mathrm{d}x$$

Clearly (i) $f(x) = F'(x)$
(ii) As $x \to -\infty$ then $F(x) = 0$ and as $x \to +\infty$, $F(x) = 1$

Worked Example 6.2

Verify that the function

$$f(x) = \begin{cases} x/2 & 0 \leqslant x \leqslant 1, \\ 1/2 & 1 < x \leqslant 2.5, \\ 0 & \text{otherwise.} \end{cases}$$

can represent a probability density function and find the corresponding distribution function.

Solution 6.2

Clearly for all x, $f(x) \geqslant 0$ and it remains to show that the total area under the curve is 1.

$$\text{Area} = \int_0^1 \frac{x}{2}\mathrm{d}x + \int_1^{2.5} \frac{1}{2}\mathrm{d}x$$

$$= \left[\frac{x^2}{4}\right]_0^1 + \left[\frac{x}{2}\right]_1^{2.5}$$

$$= (1/4) + [(5/4) - (1/2)]$$

$$= 1$$

As both conditions (i) and (ii) are satisfied, $f(x)$ can represent a probability density function.

To determine the corresponding distribution function we should note

that $P(X < 0) = 0 \Rightarrow F(x) = 0$ for $x < 0$

and $P(X > 2.5) = 1 \Rightarrow F(x) = 1$ for $x > 2.5$

In the interval $0 \leqslant x \leqslant 1$ $F(x) = \int \frac{x}{2} \mathrm{d}x$

$$= \frac{x^2}{4} + c \quad \text{Quite important!}$$

$$\text{As } F(0) = 0 \Rightarrow \frac{0}{4} + c = 0 \Rightarrow c = 0$$

In the interval $1 < x \leqslant 2.5$ $F(x) = \int \frac{1}{2} \mathrm{d}x$

$$= \frac{x}{2} + c$$

$$\text{As } F(2.5) = 1 \Rightarrow \frac{5}{4} + c = 1 \Rightarrow c = -\frac{1}{4}$$

The distribution function is therefore given by

$$F(x) = \begin{cases} 0 & x < 0, \\ \frac{x^2}{4} & 0 \leqslant x \leqslant 1, \\ \frac{x}{2} - \frac{1}{4} & 1 < x \leqslant 2.5, \\ 1 & x > 2.5. \end{cases}$$

The graphs of $f(x)$ and $F(x)$ are shown below in Fig. 6.2.

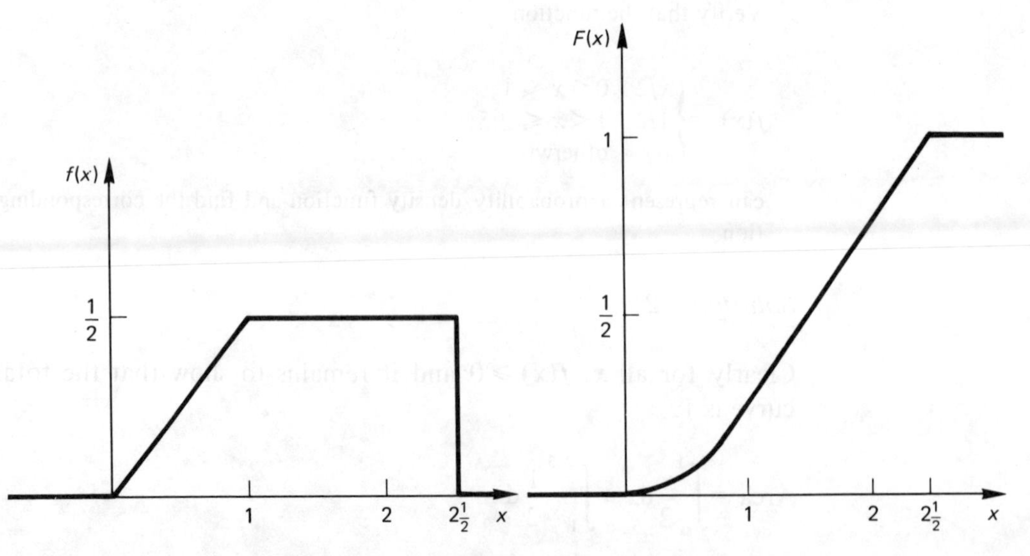

Figure 6.2

6.3 Expectation and Variance

For a discrete random variable X we defined $E(X) = \Sigma p_i x_i$. In the case of continuous distributions, integration replaces the summation sign and we define expectation of a continuous random variable X as

$$E(X) = \int_{-\infty}^{\infty} x f(x) \mathrm{d}x$$

Again we can denote this by the Greek letter μ.
In a similar manner using the previously stated definition

$\mathrm{Var}(X) = E(X^2) - (E(X))^2$

the variance of a continuous random variable X is defined as

$$\mathrm{Var}(X) = \int_{-\infty}^{\infty} x^2 f(x) \mathrm{d}x - \mu^2$$

and this can be denoted by the letter σ^2.

Worked Example 6.3

Find the mean and the variance of the probability density function defined in Worked Example 6.2 giving answers correct to 3 significant figures.

Solution 6.3

$$\mu = \int_{-\infty}^{\infty} x f(x) \, \mathrm{d}x$$

$$= \int_0^1 x \, . \, \frac{x}{2} \, \mathrm{d}x + \int_1^{2.5} x \, . \, \frac{1}{2} \, \mathrm{d}x$$

$$= \left[\frac{x^3}{6}\right]_0^1 + \left[\frac{x^2}{4}\right]_1^{2.5}$$

$$= \frac{1}{6} + \left(\frac{25}{16} - \frac{1}{4}\right)$$

$$= 71/48 = 1.48$$

$$\sigma^2 = \int_{-\infty}^{\infty} x^2 f(x) \, \mathrm{d}x - \mu^2$$

$$= \int_0^1 x^2 \, . \, \frac{x}{2} \, \mathrm{d}x + \int_1^{2.5} x^2 \, . \, \frac{1}{2} \, \mathrm{d}x - (1.479)^2$$

$$= \left[\frac{x^4}{8}\right]_0^1 + \left[\frac{x^3}{6}\right]_1^{2.5} - 2.188$$

$$= \frac{1}{8} + \left(\frac{125}{48} - \frac{1}{6}\right) \qquad - 2.188$$

$$= 2.562 - 2.188$$

$$= 0.374$$

6.4 The Continuous Uniform Distribution

The continuous random variable X has a *uniform distribution* in the interval $a \leqslant x \leqslant b$ if the probability density function is given by

$$f(x) = \begin{cases} \dfrac{1}{b-a} & a \leqslant x \leqslant b \\ 0 & \text{otherwise} \end{cases}$$

By the nature of the shape of this function it is often known as a *rectangular distribution.*

6.5 Transformations Involving the Uniform Distribution

Sometimes probability density functions are dependent on an underlying uniform distribution. That is, a random variable $Y = g(X)$ where X is rectangularly distributed in a given interval and for each value of Y there corresponds one and only one value of X.

If $Y = g(X)$

then $P(y \leqslant Y \leqslant y + \delta y) = P(x \leqslant X \leqslant x + \delta x)$ (see Fig. 6.3.)

But $P(y \leqslant Y \leqslant y + \delta y) = f(y)\,|\,\delta y\,|$ could be negative

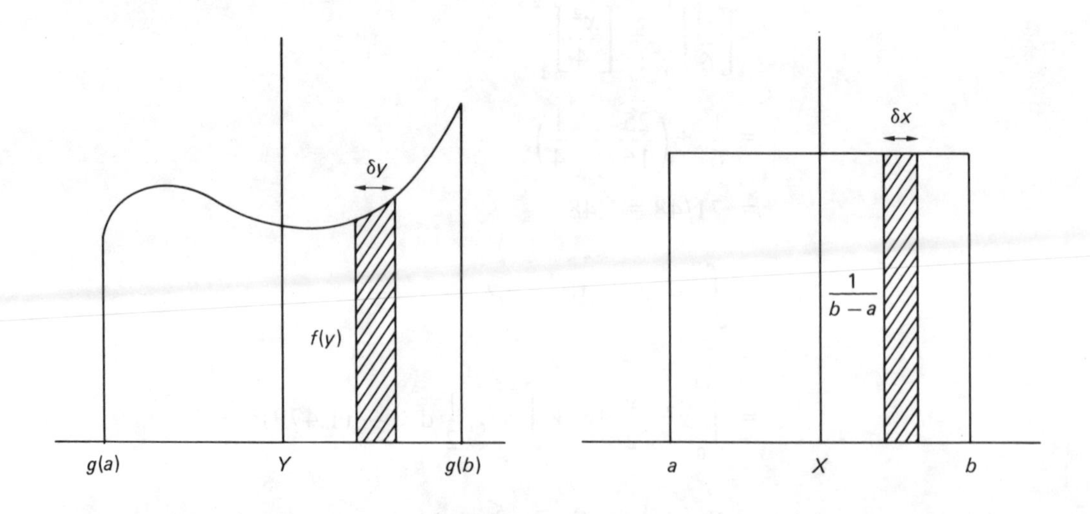

Figure 6.3

$$\text{and} \quad P(x \leqslant X \leqslant x + \delta x) = \frac{1}{b-a}\delta x$$

$$\Rightarrow f(y)\,|\delta y| = \frac{1}{b-a}\delta x$$

$$\Rightarrow f(y) = \frac{1}{b-a}\left|\frac{\delta x}{\delta y}\right|$$

$$\text{Now as } \delta x, \delta y \to 0 \ f(y) = \begin{cases} \dfrac{1}{b-a}\left|\dfrac{\mathrm{d}x}{\mathrm{d}y}\right| & g(a) \leqslant y \leqslant g(b) \\ 0 & \text{otherwise} \end{cases}$$

It is important to appreciate that there should be only one value of X for each value of Y. If this is not the case an adjustment is necessary as in Worked Example 6.4.

Worked Example 6.4*

A chord is drawn in a random direction from a point P on the circumference of a circle of radius r. If α is the angle between the chord and the diameter through P and α is distributed uniformly between $-\pi/2$ and $+\pi/2$, show that the probability density function of the length L of the chord is given by

$$y(l) = \begin{cases} \dfrac{2}{\pi(4r^2 - l^2)^{\frac{1}{2}}} & 0 \leqslant l \leqslant 2r \\ 0 & \text{otherwise} \end{cases}$$

*Solution 6.4**

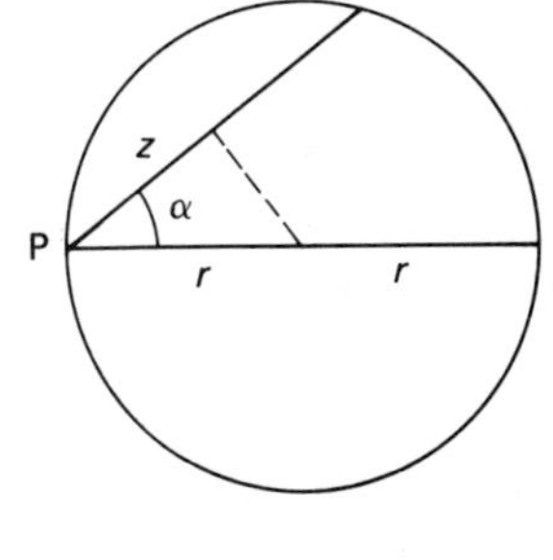

Figure 6.4

α is uniformly distributed

$$f(\alpha) = \begin{cases} \dfrac{1}{\pi} & -\dfrac{\pi}{2} \leqslant \alpha \leqslant \dfrac{\pi}{2} \\ 0 & \text{otherwise} \end{cases}$$

$$z = r\cos\alpha$$

$$\Rightarrow l = 2z = 2r\cos\alpha$$

The distribution of the random variable L which lies between 0 and $2r$ is

$$L = 2r\cos\alpha$$

Each L can be generated from not one but *two* possible values of α, therefore

$$y(l) = \begin{cases} \dfrac{2}{\dfrac{\pi}{2} - \left(\dfrac{-\pi}{2}\right)}\left|\dfrac{\mathrm{d}\alpha}{\mathrm{d}l}\right| & 0 \leqslant l \leqslant 2r \\ 0 & \text{otherwise} \end{cases}$$

$$\text{Now } \frac{\mathrm{d}l}{\mathrm{d}\alpha} = -2r\sin\alpha$$

$$= -2r\sqrt{\sin^2\alpha}$$

$$= -2r\sqrt{1 - \cos^2\alpha}$$

$$= -2r\sqrt{1-\frac{l^2}{4r^2}}$$

$$= -\sqrt{4r^2 - l^2}$$

So $$y(l) = \begin{cases} \dfrac{2}{\pi(4r^2 - l^2)^{\frac{1}{2}}} & 0 \leqslant l \leqslant 2r \\ 0 & \text{otherwise} \end{cases}$$

Worked Example 6.5

A continuous random variable X has a probability density function defined as follows

$$f(x) = \begin{cases} c(4-x^2) & -2 \leqslant x \leqslant 2 \\ 0 & \text{Otherwise} \end{cases}$$

Find (i) the value of c
(ii) the mean
(iii) the variance
(iv) the cumulative distribution function.

Solution 6.5

(i) $$\int_{-2}^{2} c(4-x^2)\,dx = 1$$

$$c\left[4x - \frac{x^3}{3}\right]_{-2}^{2} = 1$$

$$\Rightarrow c\left[\left(8 - \frac{8}{3}\right) - \left(-8 + \frac{8}{3}\right)\right] = 1$$

$$\frac{32c}{3} = 1 \Rightarrow c = 3/32$$

(ii) $$\mu = \frac{3}{32}\int_{-2}^{2} x\,.(4-x^2)dx$$

$$= \frac{3}{32}\left[2x^2 - \frac{x^4}{4}\right]_{-2}^{2}$$

$$= \frac{3}{32}\,[(8-4)-(8-4)]$$

$$= 0$$

(iii) $$\sigma^2 = \frac{3}{32}\int_{-2}^{2} x^2\,.(4-x^2)dx - 0^2$$

$$= \frac{3}{32}\left[\frac{4x^3}{3} - \frac{x^5}{5}\right]_{-2}^{2}$$

$$= \frac{3}{32}\left[\left(\frac{32}{3} - \frac{32}{5}\right) - \left(\frac{-32}{3} - \frac{-32}{5}\right)\right]$$

$$= \frac{3}{32} \times \frac{128}{15}$$

$$= 4/5$$

(iv) $P(X < -2) = 0$ so for $x < -2$, $F(x) = 0$

$P(X > 2) = 1$ so for $x > 2$, $F(x) = 1$

In the range $-2 \leqslant x \leqslant 2$, $F(x) = \int \frac{3}{32}(4 - x^2)\mathrm{d}x$

$$= \frac{3}{32}\left(4x - \frac{x^3}{3}\right) + c$$

Now when $x = -2$, $F(x) = 0$

$$\Rightarrow \frac{3}{32}\left[-8 - \left(\frac{-8}{3}\right)\right] + c = 0$$

$$\Rightarrow \quad c = 1/2$$

So

$$F(x) = \begin{cases} 0 & x < -2 \\ \frac{3}{32}\left(4x - \frac{x^3}{3}\right) + \frac{1}{2} & -2 \leqslant x \leqslant 2 \\ 1 & x > 2 \end{cases}$$

Worked Example 6.6

A probability density function is defined as

$$f(x) = \begin{cases} \frac{3}{4}x(2 - x) & 0 \leqslant x \leqslant 2 \\ 0 & \text{otherwise} \end{cases}$$

Find $P(|X - \mu| < \sigma)$

Solution 6.6

Before we can attempt the solution we must first understand what is meant by the probability statement '$P(|X - \mu| < \sigma)$'. $X - \mu$ is contained between two vertical lines which mathematically means whatever is the result of subtracting μ from X, always take it as positive. It is known as the *modulus* of $X - \mu$.

The statement $|X - \mu| < \sigma$ can therefore be written as

$-\sigma < X - \mu < \sigma$

or adding μ to each term $\mu - \sigma < X < \mu + \sigma$

and so we require the area between $\mu - \sigma$ and $\mu + \sigma$.

Mean $\qquad \mu = \int_0^2 \frac{3}{4} \cdot x^2(2 - x)\,\mathrm{d}x$

$$= \frac{3}{4}\left[\frac{2x^3}{3} - \frac{x^4}{4}\right]_0^2$$

$$= \frac{3}{4}\left[\frac{16}{3} - 4\right] = 1$$

Variance $\quad \sigma^2 = \int_0^2 \frac{3}{4} \cdot x^3(2 - x)\,dx - 1^2$

$$= \frac{3}{4}\left[\frac{2x^4}{4} - \frac{x^5}{5}\right]_0^2 - 1$$

$$= \frac{3}{4}\left(8 - \frac{32}{5}\right) - 1$$

$$= 1/5$$

So the standard deviation $= 0.447$

So $P(|X - \mu| < \sigma) = \int_{1-0.447}^{1+0.447} \frac{3}{4} . x(2 - x)\,dx$

$$= \frac{3}{4}\left[x^2 - \frac{x^3}{3}\right]_{0.553}^{1.447}$$

$$= \frac{3}{4}[1.084 - 0.249]$$

$$= 0.626$$

Worked Example 6.7

A random variable X has a cumulative distribution function defined as

$$F(x) = \begin{cases} 0 & x < 0 \\ mx^n & 0 \leqslant x \leqslant 2 \\ 1 & x > 2 \end{cases}$$

If the mean of the distribution is 2/3, find m and n.

Solution 6.7

$$f(x) = F'(x)$$

$$= \begin{cases} mnx^{n-1} & 0 \leqslant x \leqslant 2 \\ 0 & \text{otherwise} \end{cases}$$

Now $\quad \mu = \int_0^2 xmnx^{n-1}\,dx = \frac{2}{3}$

$$\Rightarrow \int_0^2 mnx^n\,dx = \frac{2}{3}$$

$$mn\left[\frac{x^{n+1}}{n+1}\right]_0^2 = \frac{2}{3}$$

$$\frac{mn}{n+1}\left[2^{n+1}-0\right]=\frac{2}{3}$$

$$\Rightarrow \quad \frac{mn}{n+1}\times 2^{n+1}=\frac{2}{3} \qquad \text{(i)}$$

and $$\int_0^2 mnx^{n-1}\,\mathrm{d}x = 1$$

$$\Rightarrow \quad \frac{mn}{n}\left[x^n\right]_0^2 = 1$$

$$m\,.\,2^n = 1 \qquad \text{(ii)}$$

Substituting into (i)

$$\frac{n\,.\,2}{n+1}=\frac{2}{3}$$

$$3n = n+1$$

$$\Rightarrow n = 1/2$$

Substituting into (ii) $\quad m\sqrt{2} = 1$

$$\Rightarrow \quad m = 1/\sqrt{2}$$

Worked Example 6.8*

A number of students are asked to draw a circle of radius 8 cm using a pair of compasses. If the radius R is rectangularly distributed in the interval 7.9 cm to 8.1 cm, find the probability density function of the area A and hence find the proportion of students you would expect to have drawn a circle less than 200 cm^2 in area.

*Solution 6.8**

R is rectangularly distributed

$$\Rightarrow \quad f(r)=\begin{cases}5 & 7.9\leqslant r\leqslant 8.1\\ 0 & \text{otherwise}\end{cases}$$

The probability density function $A = \pi R^2$

$$\text{So } g(a)=\begin{cases}\dfrac{1}{0.2}\left|\dfrac{\mathrm{d}r}{\mathrm{d}a}\right| & 62.41\pi\leqslant a\leqslant 65.61\pi\\ 0 & \text{otherwise}\end{cases}$$

$$\text{Now } \frac{\mathrm{d}A}{\mathrm{d}R} = 2\pi R$$

$$= 2\pi\sqrt{\frac{A}{\pi}}$$

$$\text{So } g(a)=\begin{cases}\dfrac{5}{2\sqrt{\pi a}} & 62.41\pi\leqslant a\leqslant 65.61\pi\\ 0 & \text{otherwise}\end{cases}$$

The proportion of students who draw a circle of area less than 200 cm

$$= \int_{62.41\pi}^{200} \frac{5}{2\sqrt{\pi a}} \, da$$

$$= \frac{5}{2\sqrt{\pi}} \int_{62.41\pi}^{200} a^{-\frac{1}{2}} \, da$$

$$= \frac{5}{2\sqrt{\pi}} [2\sqrt{a}]_{62.41\pi}^{200}$$

$$= \frac{50\sqrt{2}}{\sqrt{\pi}} - 5 \times 7.9$$

$$= 39.90 - 39.5$$

$$= 0.4$$

Worked Example 6.9

The quality of an animal feedstuff depends both upon the raw materials used and the production process. One measure of this quality is the Nutritional Index, which varies between 0 and 1. A particular mill produces an animal feedstuff in batches of constant size, and the Nutritional Index of any batch may be considered to be an observation on a random variable X having probability density function

$$f(x) = \begin{cases} kx(1-x)^2 & 0 \leqslant x \leqslant 1 \\ 0 & \text{otherwise} \end{cases}$$

(a) Show that $k = 12$, and sketch $f(x)$.
(b) Calculate $P(X < 0.25)$.
(c) Determine the mean of X.
(d) Batches of this mill's feedstuff may be sold for £500 each if the index is 0.8 or more, and £350 otherwise. The cost of producing a batch is £300. What is the expected profit per batch?

(AEB 1983)

Solution 6.9

(a) $\int_0^1 kx(1-x)^2 \, dx = 1$

$$k\left[\frac{x^2}{2} - \frac{2x^3}{3} + \frac{x^4}{4}\right]_0^1 = 1$$

$$k\left[\frac{1}{2} - \frac{2}{3} + \frac{1}{4}\right] = 1$$

$$\frac{k}{12} = 1 \Rightarrow k = 12$$

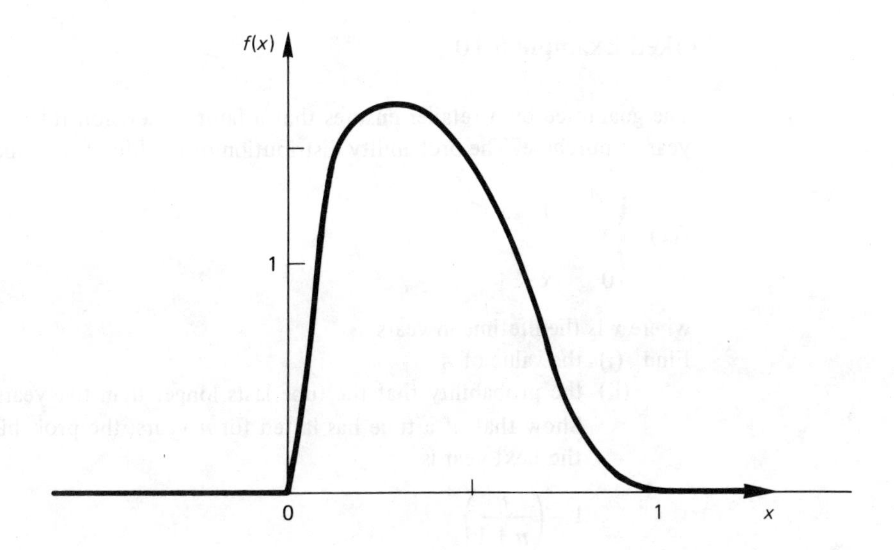

Figure 6.5

(b) $$P(X < 0.25) = \int_0^{0.25} 12(x^3 - 2x^2 + x)\,dx$$

$$= 12\left[\frac{x^4}{4} - \frac{2x^3}{3} + \frac{x^2}{2}\right]_0^{0.25}$$

$$= 12(0.0010 - 0.0104 + 0.0312)$$

$$= 0.262$$

(c) $$\mu = \int_0^1 12(x^4 - 2x^3 + x^2)\,dx$$

$$= 12\left[\frac{x^5}{5} - \frac{x^4}{2} + \frac{x^3}{3}\right]_0^1$$

$$= 12\left(\frac{1}{5} - \frac{1}{2} + \frac{1}{3}\right) = \frac{2}{5}$$

(d) $$P(X \geqslant 0.8) = 1 - P(X < 0.8)$$

$$= 1 - 12\left[\frac{x^4}{4} - \frac{2x^3}{3} + \frac{x^2}{2}\right]_0^{0.8}$$

$$= 1 - 12(0.1024 - 0.3413 + 0.32)$$

$$= 0.0272$$

Expected profit/batch $$= £(\underbrace{350 \times 0.9728}_{P(X<0.8)} + \underbrace{500 \times 0.0272}_{P(X>0.8)} - 300)$$

$$= £(340.48 + 13.60 - 300.00)$$

$$= £54.08$$

Worked Example 6.10

The guarantee by a retailer ensures that a faulty television tube is replaced within the first year of purchase. The probability distribution of the life of the tube can be modelled by

$$f(x) = \begin{cases} \dfrac{A}{x^4} & 1 \leqslant x \leqslant \infty \\ 0 & x < 1 \end{cases}$$

where x is the lifetime in years.

Find (i) the value of A;

(ii) the probability that the tube lasts longer than five years.

Show that, if a tube has lasted for n years, the probability that it will fail during the next year is

$$1 - \left(\frac{n}{n+1}\right)^3$$

(SUJB)

Solution 6.10

(i) $\displaystyle\int_1^\infty \frac{A}{x^4}\,dx = 1$

$$A\left[\frac{x^{-3}}{-3}\right]_1^\infty = 1$$

$$\Rightarrow A(0 - (-1/3)) = 1 \Rightarrow A = 3$$

(ii) P(Tube lasts longer than 5 years)

$$= \int_5^\infty \frac{3}{x^4}\,dx = \left[\frac{x^{-3}}{-1}\right]_5^\infty$$

$$= 1/125$$

P(Tube fails during $(n + 1)$th year/Survived up to nth year)

$= 1 - P$(Tube survives $(n + 1)$th year/Survived up to nth year)

$$= 1 - \frac{P(\text{Tube survives to at least } (n+1)\text{th year})}{P(\text{Tube survives up to } n\text{th year})}$$

$$= 1 - \frac{\displaystyle\int_{n+1}^\infty \frac{3}{x^4}\,dx}{\displaystyle\int_n^\infty \frac{3}{x^4}\,dx} = 1 - \frac{\left[\dfrac{-1}{x^3}\right]_{n+1}^\infty}{\left[\dfrac{-1}{x^3}\right]_n^\infty}$$

$$= 1 - \frac{\left[\dfrac{1}{(n+1)^3}\right]}{\dfrac{1}{n^3}}$$

$$= 1 - \left(\frac{n}{n+1}\right)^3$$

6.6 Exercises

Exercise 6.1

The probability density function of a random variable X is given by

$$f(x) = \begin{cases} k(1-x)(x-4) & 1 \leqslant x \leqslant 4 \\ 0 & \text{otherwise} \end{cases}$$

Find the value of k and the mean and variance of the distribution. Find also the probability that X is greater than 2.

Exercise 6.2

A continuous random variable X takes values in the interval $x > 0$. If the probability density function is defined as

$$f(x) = \begin{cases} \frac{1}{4}x & 0 \leqslant x \leqslant 1 \\ \frac{1}{4x^n} & x > 1 \end{cases}$$

Find n.

Exercise 6.3

A ladder of length 6 m rests on horizontal ground a distance X from a vertical wall. Given that the distance X from the base of the wall is uniformly distributed in the interval 1.0 to 2.0 m, find the probability that the height of the ladder against the wall exceeds 5.75 m.

Exercise 6.4*

A boy tries his model boat out on a circular pond of radius 20 m. Given that he places the boat at the edge of the pond and that he is equally likely to point the boat towards any other point on the circumference of the pond, show that the probability density function of the distance d that the boat travels is

$\dfrac{2}{\pi\sqrt{1600-d^2}}$ for values of d between 0 and 40 and zero otherwise.

If the boat travels at 0.5 m/sec, find the expected time elapse before the boat reaches the side again.

Exercise 6.5

A continuous random variable X has a probability density function given by

$$f(x) = \begin{cases} ax & 0 \leqslant x \leqslant 2 \\ 2a & 2 < x \leqslant 3 \\ 0 & \text{otherwise} \end{cases}$$

Calculate the value of a, and the mean and median of the distribution. Calculate the cumulative distribution $F(x)$ and give a rough sketch of $f(x)$ and $F(x)$.

(UCLES)

Exercise 6.6

The cumulative distribution function for a continuous random variable is given by

$$F(x) = \begin{cases} 0 & x < -2 \\ k\left(4x - \dfrac{1x^3}{3} + \dfrac{16}{3}\right) & -2 \leqslant x \leqslant 2 \\ 1 & x > 2 \end{cases}$$

Find

(i) the value of k,
(ii) $P(-1 < X < 1)$,
(iii) the probability density function for X,
(iv) the mean and variance of X.

(UCLES)

Exercise 6.7

A continuous variable is distributed with frequency-density z where $z = 2x$ if $0 \leqslant x \leqslant 2$ and $z = 6 - x$ if $2 \leqslant x \leqslant 6$.

Draw a sketch graph of z against x for $0 \leqslant x \leqslant 6$ and show that the probability density function is $\dfrac{x}{6}$ if $0 \leqslant x \leqslant 2$ and $\dfrac{6-x}{12}$ if $2 < x \leqslant 6$.

Hence find

(i) the mean value of x;
(ii) the standard deviation;
(iii) the median value.

(SUJB)

Exercise 6.8

The continuous random variable X has probability density function f given by

$$f(x) = \begin{cases} k & 0 \leqslant x \leqslant 2 \\ k(3 - x) & 2 < x \leqslant 3 \\ 0 & \text{otherwise} \end{cases}$$

(i) Find the value of k.
(ii) Hence evaluate $E(X)$ and the median of X.
(iii) Prove that σ, the standard deviation of X, is 0.75 correct to 2 decimal places.
(iv) Denoting $E(X)$ by μ, find $P(X < (\mu - \sigma))$.

(UCLES)

Exercise 6.9*

X is a continuous random variable with probability density function given by

$$f(x) = \begin{cases} ke^{-2x} & x \geqslant 0 \\ 0 & x < 0 \end{cases}$$

(i) Prove that $k = 2$.
(ii) Calculate $E(X)$ and $\text{Var}(X)$.
(iii) Prove that the median m is equal to $(\ln 2)/2$. (UCLES)

Exercise 6.10

The lifetime in years of a television tube of a certain make is a random variable T and its probability density function $f(t)$ is given by

$$f(t) = \begin{cases} Ae^{-kt} & 0 \leqslant t \leqslant \infty \ (k > 0) \\ 0 & \text{otherwise} \end{cases}$$

Obtain A in terms of k.

(i) If the manufacturer, after some research, finds that out of 1000 such tubes 371 failed within the first two years of use, estimate the value of k.
(ii) If two such tubes are bought, what is the probability that one fails within its first year and the other lasts longer than six years. (SUJB)

Exercise 6.11*

The continuous random variable X has probability density function f, where

$$f(x) = \begin{cases} cx(2-x) & 0 \leqslant x \leqslant 2 \\ 0 & \text{otherwise} \end{cases}$$

(i) Find the value of c.
(ii) Show that the variance of X is 0.2.
(iii) Determine the cumulative distribution function of X.
(iv) Given that

$$y = \frac{x}{4-x}$$

determine the range R of the values of y corresponding to values of x ranging from 0 to 2. Show that the probability density function g of

$$Y = \frac{X}{4-X}$$

is given by

$$g(y) = \frac{ky(1-y)}{(1+y)^4} \quad \text{for } y \text{ in } R$$

$$g(y) = 0 \quad \text{otherwise}$$

and write down the value of the constant k. (WJEC)

7 The Normal Distribution

7.1 Properties of the Normal Distribution

In Chapter 5 we looked at two important discrete probability distributions and their relevant applications. We will now consider perhaps the most important of the continuous distributions – the *Normal distribution*.

A significant number of distributions start with a low frequency, rise to a maximum towards the centre of the distribution and then fall away again giving a somewhat symmetrical pattern to the distribution. On drawing the frequency curve the distribution takes on the appearance of the cross-section of a bell. It is this shape which characterises the normal distribution.

The Normal distribution is a probability density function and has the following properties:

1. The function is defined as:

$$f(x) = \frac{1}{\sigma\sqrt{2\pi}} e^{-\frac{1}{2}\left(\frac{x-\mu}{\sigma}\right)^2} \quad -\infty < x < \infty$$

   (i) The function is totally defined by its mean μ and standard deviation σ and is sometimes denoted as $N(\mu, \sigma^2)$.
   (ii) For all $x, f(x) > 0$.
   (iii) $\int_{-\infty}^{\infty} f(x)\,dx = 1$

2. It is a symmetrical, bell shaped function with the mean, median and mode coinciding at the maximum value of the function.
3. It is asymptotic to the x-axis. i.e. as $x \to \pm\infty, f(x) \to 0$.
4. Although the distribution is defined for values of x from $-\infty$ to $+\infty$, all but 0.3% of the area lies between $\pm 3\sigma$ of the mean. As a guide the diagram (Fig. 7.1) illustrates relevant areas and associated probabilities of the distribution.

7.2 Standardisation

It is clear that there are an infinite number of such functions each uniquely defined by a mean and standard deviation. The calculation of areas underneath any curve can prove to be quite laborious but fortunately for the reader any area can be determined from statistical tables (see Table 16.1, page 250).

In order to use Table 16.1 any random variable X which is normally distributed is transformed using the transformation

$$Z = \frac{X - \mu}{\sigma}$$

The Worked Example 4.13 shows that by using this transformation, a distribution of mean μ and standard deviation σ reduces to a distribution with mean 0 and standard deviation 1.

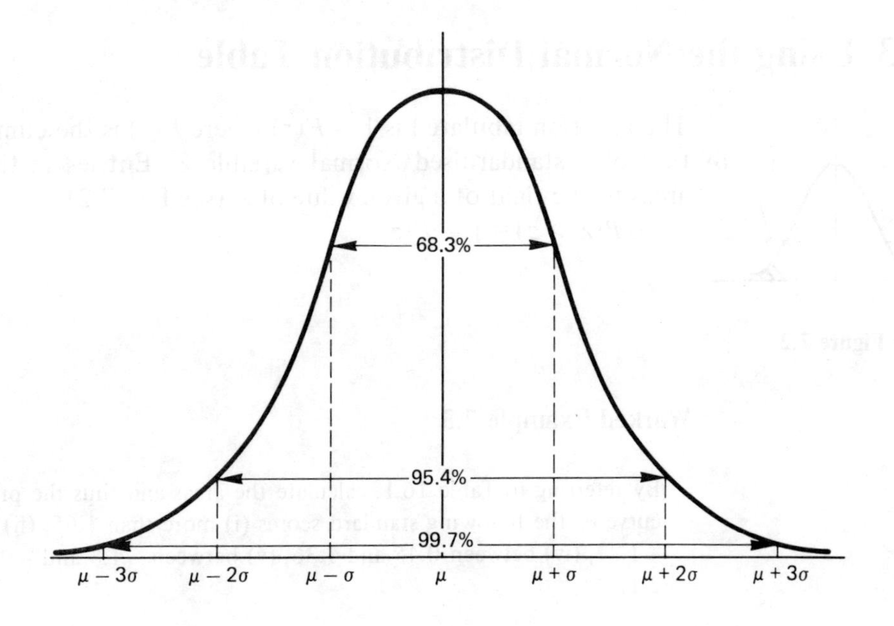

Figure 7.1 The Normal probability distribution curve

Z is distributed normally and can be written $N(0, 1^2)$ and any value z from this distribution is called a *standard score*.

Worked Example 7.1

Given a mean of 150 and a standard deviation of 9, calculate the standard score of the value 138.

Solution 7.1

The standard score, $z = \dfrac{138 - 150}{9} = -1.33$

N.B. It is unnecessary to give any greater degree of accuracy in the answer.

Worked Example 7.2

Given a standard score of 1.2, a mean of 32.2 and a standard deviation of 4.5, calculate the original value.

Solution 7.2

By changing the subject of the original formula we arrive at

$$x = \mu + z\sigma$$
$$= 32.2 + 1.2 \times 4.5$$
$$= 37.6$$

7.3 Using the Normal Distribution Table

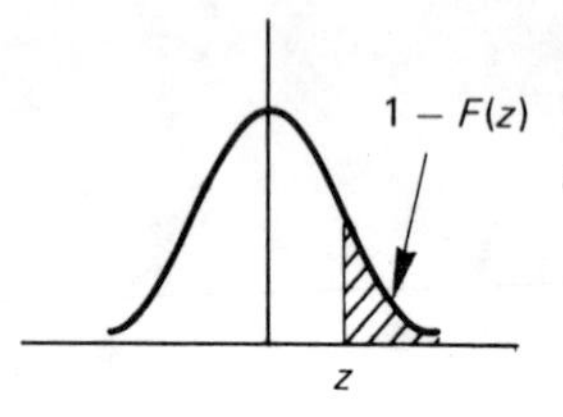

Figure 7.2

The function tabulated is $1 - F(z)$ where $F(z)$ is the cumulative distribution function of a standardised Normal variable Z. Entries in the body of the table give areas to the right of a given value of Z (see Fig. 7.2).

So $P(Z > z) = 1 - F(z)$.

Worked Example 7.3

By referring to Table 16.1, calculate the areas and thus the probabilities under the Normal curve of the following standard scores (i) more than 1.65, (ii) less than 2.0, (iii) more than −1.27, (iv) between 0.45 and 1.36, (v) between −1.8 and −0.23, (vi) between −0.75 and 2.1.

Solution 7.3

(i) The area to the right of $z = 1.65$ is required

$P(Z > 1.65) = 0.0495$

(ii) $P(Z < 2.0) = 1 - P(Z > 2.0)$
$= 1 - 0.02275$
$= 0.97725$

(iii) The table only shows positive values of the standardised scores but by using the symmetry property of the distribution, the area required to the right of −1.27 is the same as the area to the left of 1.27.

$P(Z > -1.27) = P(Z < 1.27)$
$= 1 - P(Z > 1.27)$
$= 1 - 0.1020$
$= 0.8980$

(iv) $P(0.45 < Z < 1.36) = P(Z > 0.45) - P(Z > 1.36)$
$= 0.3264 - 0.0869$
$= 0.2395$

(v) By using the symmetry property of the distribution

$P(-1.8 < Z < -0.23) = P(0.23 < Z < 1.8)$
$= P(Z > 0.23) - P(Z > 1.8)$
$= 0.4090 - 0.0359$
$= 0.3731$

(vi) $P(-0.75 < Z < 2.1) = 1 - P(Z < -0.75 \text{ or } Z > 2.1)$
$= 1 - [P(Z < -0.75) + P(Z > 2.1)]$
$= 1 - [P(Z > 0.75) + P(Z > 2.1)]$
$= 1 - [0.2266 + 0.01786]$
$= 1 - 0.24446$
$= 0.75554$

Worked Example 7.4

Calculate the standard score values z defined by the following areas

(i) the area to the right of z is 0.2061,
(ii) the area to the left of z is 0.0606,
(iii) the area between −2 and z is 0.9665.

Solution 7.4

(i) $P(Z > z) = 0.2061 \quad \Rightarrow z = 0.82$ from the body of the table.

(ii) As the area to the right of z exceeds 0.5, z is negative. By symmetry the area to the left of a negative value is the same as the area to the left of its positive counterpart.

$P(Z < z) = 0.0606$ is the same as $P(Z > z') = 0.0606$

where $\quad z' = -z$

From the table, $z' = 1.55 \Rightarrow z = -1.55$

(iii) $P(-2 < Z < z) \quad = P(Z > -2) - P(Z > z)$

and $P(Z > -2) \quad = 1 - P(Z > 2) = 0.97725$

So $0.97725 - P(Z > z) = 0.9665$

$P(Z > z) = 0.97725 - 0.9665 = 0.01075$

From the table $z = 2.3$

Worked Example 7.5

In an external examination, the scores are normally distributed with mean 120 and standard deviation 15. If a score of 100 is required to pass the examination, what is the probability that a candidate chosen at random has failed the examination?

Solution 7.5

To find the required probability we first of all have to standardise 100.

$$z = \frac{100 - 120}{15} = -1.33$$

The area to the left of -1.33 will now give the answer.
Therefore in the usual probability notation:

$$\begin{aligned} P(\text{A candidate fails}) &= P(\text{He scores less than } 100) \\ &= P(Z < -1.33) \\ &= P(Z > 1.33) \\ &= 0.0918 \end{aligned}$$

7.4 Calculating Theoretical Frequencies

Given a reasonably symmetrical distribution of data it is possible to fit a theoretical Normal distribution. This is usually done by estimating the mean and standard deviation from the sample data.

Worked Example 7.6

The marks obtained by 500 candidates in an examination are given in Table 7.1.

Table 7.1

Marks (central value)	15	25	35	45	55	65	75	85	95
Frequency	10	35	98	120	110	80	32	9	6

Fit a Normal distribution to this data and hence estimate the percentage of candidates with a mark under 35.

Solution 7.6

Using the methods of Chapter 2, we find the mean is 49.68 and the standard deviation 15.69. The theoretical Normal distribution that fits this data is $N(49.68, 15.69^2)$.

Standardising 35 we have

$$z = \frac{35 - 49.68}{15.69} = -0.94$$

The percentage of candidates with a mark under 35 is equal to the area to the left of the standardised score of -0.94. From the table this gives 17.36%.

N. B. Sheppard's correction which was briefly mentioned in Section 2.18 has not been used here and it is left to the reader to try the question using the corrected standard deviation.

Differences between data and theoretical distributions are to be expected and in Chapter 12 we investigate how well the theoretical distribution fits the related data.

7.5 Approximations to the Binomial and Poisson Distributions

In Chapter 5 we have seen that if n is large, calculations involving the binomial distribution become very tedious. However providing that both np and nq are greater than 5, the Normal distribution is a good approximation to the discrete distribution. Since the values in a binomial distribution are integral, a continuity correction factor is required when using the Normal distribution as an approximation. As the mean of the binomial distribution is np and its standard deviation $\sqrt{npq}$, the standard score z is calculated from

$$z = \frac{x \pm 0.5 - np}{\sqrt{npq}}$$

Worked Example 7.7

On a multiple choice question paper there are 50 question with five alternative answers to each. There is only one correct answer to each question. Find, by a suitable method, the probability of exceeding 15 by random guessing.

Solution 7.7

The binomial distribution would be rather tedious for this question and as both np and nq are greater than 5 we should use the Normal approximation to this distribution.

$$\mu = np = 50 \,.\, 0.2 = 10$$
$$\sigma = \sqrt{npq} = \sqrt{50 \times 0.2 \times 0.8} = 2.83$$

The two distributions, binomial and theoretical Normal are shown in Fig. 7.3. For a value exceeding 15, we require to sum the probabilities for marks 16, 17, . . . and so on. The value to standardise is $15 + 0.5 = 15.5$

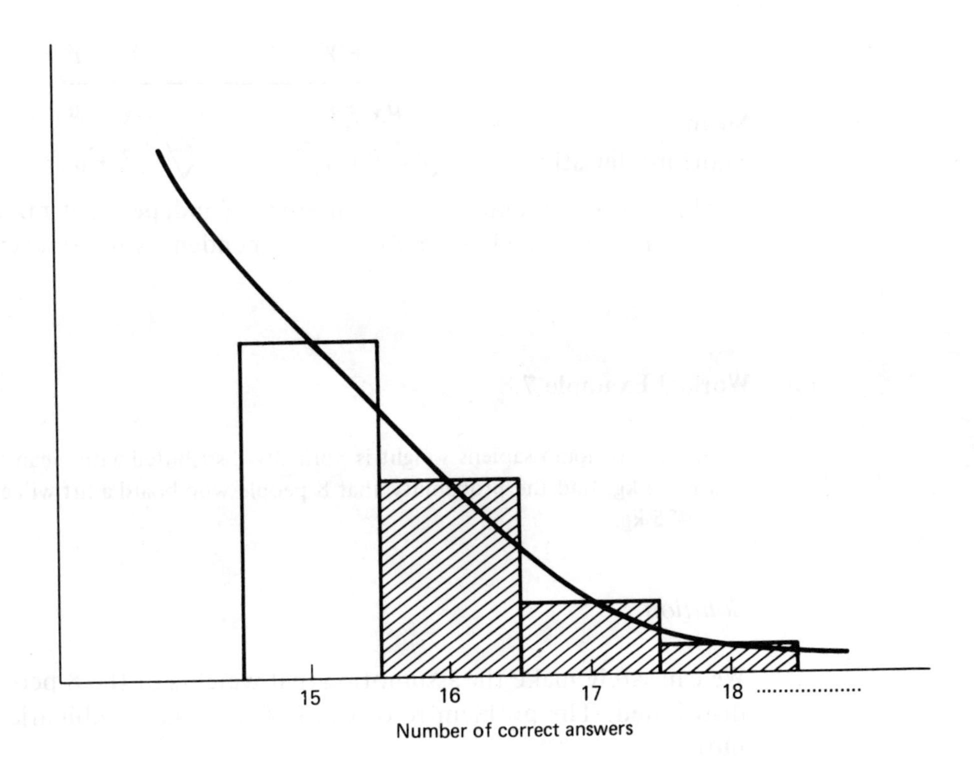

Figure 7.3

Standardising 15.5

$$z = \frac{15.5 - 10}{2.83} = 1.94$$

Therefore P(Candidate exceeds a mark of 15 by random guesses)

$= P(Z > 1.94)$
$= 0.0262$

As there is a relation between the binomial and Poisson distributions, so there is also a relation between the Poisson and Normal distributions. Providing that the mean m of the Poisson distribution is greater than about 15 and the tails are excluded, the Normal distribution can be used as an adequate approximation and the standard score is calculated from

$$Z = \frac{x \pm 0.5 - m}{m^{\frac{1}{2}}}$$

Again note the continuity correction factor.

7.6 Combining Distributions

In Chapter 4 we have shown that providing random variables are independent we can combine two variables, X and Y, together to form a new random variable. If μ_X, μ_Y and σ_X, σ_Y are the respective means and standard deviations of X and Y, the mean and standard deviation of the distributions $X + Y$ and $X - Y$ are

	$X + Y$	$X - Y$
Mean	$\mu_X + \mu_Y$	$\mu_X - \mu_Y$
Standard deviation	$\sqrt{\sigma_X^2 + \sigma_Y^2}$	$\sqrt{\sigma_X^2 + \sigma_Y^2}$

This can be extended to any number of independent random variables and in particular is useful when combining independent Normal distributions.

Worked Example 7.8

Given that homo sapiens weight is normally distributed with mean 55 kg and standard deviation 5 kg, find the probability that 8 people who board a lift will exceed the safety weight of 475 kg.

Solution 7.8

We can safely make the assumption that weights of the 8 people are independently distributed. The problem resolves itself into the combination of 8 normal distributions.

The mean of the combined distribution $= 8 \times 55 = 440$ kg

The standard deviation $= \underbrace{\sqrt{5^2 + 5^2 \ldots 5^2}}_{8 \text{ times}} = 14.14$ kg

Standardising 475

$$z = \frac{475 - 440}{14.14} = 2.48$$

P(The weight of 8 people will exceed 475 kg)

$$= P(Z > 2.48) = 0.0066$$

Worked Example 7.9

A machine is used to package 25 kg of potatoes. Given that the weights are normally distributed with standard deviation of 0.04 kg, what should be the setting of the machine in order that 95% of the bags exceed the nominal weight of 25 kg?

Solution 7.9

The unknown to be found is the mean of the Normal distribution. We require that 95% of all bags have to exceed 25 kg. Therefore the appropriate z-value such that the area to the right (Fig. 7.4) is 0.95 is -1.645.
Using the standardisation formula and making the mean the subject,

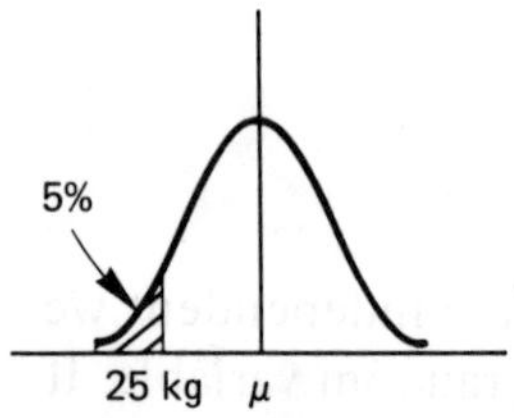

$$\mu = x - \sigma . z$$
$$= 25 - 0.04 . (-1.645)$$
$$= 25.066 \text{ kg}$$

5%
25 kg μ

Figure 7.4

Worked Example 7.10

The weights of a set of castings are normally distributed. 10.75% of them weigh over 25 kg and 4.99% weigh under 15 kg. Calculate the mean weight and the standard deviation of the distribution. (SUJB)

Solution 7.10

As we have two unknowns we shall solve the problem by using simultaneous equations. The two equations are generated from the standardisation formula but first we have to find the two standardised z-values.

$P(X > 25) = 0.1075 \Rightarrow z_1 = 1.24$

$P(X < 15) = 0.0499 \Rightarrow z_2 = -1.645$

Using both these values in the standardisation formula

$$1.24 = \frac{25 - \mu}{\sigma}$$

$$\mu = 25 - 1.24\sigma \ldots\ldots \text{(i)}$$

and $$-1.645 = \frac{15 - \mu}{\sigma}$$

$$\mu = 15 + 1.645\sigma \quad \text{(ii)}$$

Equating (i) and (ii)

$$25 - 1.24\sigma = 15 + 1.645\sigma$$

$$10 = 2.885\sigma$$

$$\Rightarrow \sigma = 3.466$$

Substituting into (i) $\mu = 25 - 4.298$

$= 20.702$

Worked Example 7.11

A population is normally distributed with a mean of 7 and a standard deviation of 0.6. Find the probability that

(i) one observation chosen at random exceeds 7.3,
(ii) that each of four observations chosen at random is less than 7.3.

Given that a random observation is less than 7.3, what is the probability that it is less than 7?

Solution 7.11

(i) Standardising 7.3

$$z = \frac{7.3 - 7}{0.6} = 0.5$$

Therefore $P(\text{observation} > 7.3) = P(Z > 0.5)$
$= 0.3085$

(ii) $P(\text{four observations} < 7.3) = (1 - P(\text{observation} > 7.3))^4$
$= (0.6915)^4$
$= 0.229$

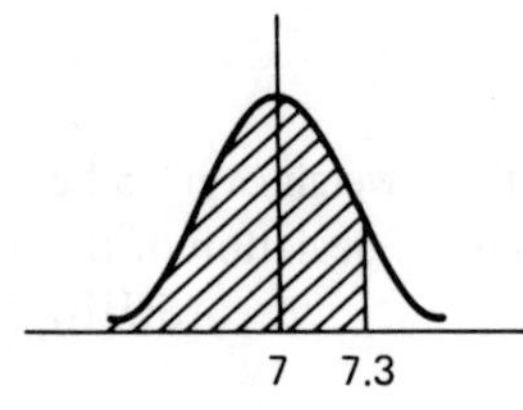

Figure 7.5

The final part of this question involves conditional probability and the restricted sample space of area 0.6915 is shown shaded in Fig. 7.5. As 7 is the mean of the distribution, the probability that the random observation is less than 7 is 0.5.

P(observation is less than 7/the observation is less than 7.3)

$$= \frac{P(\text{observation} < 7)}{P(\text{observation} < 7.3)}$$

$$= \frac{0.5}{0.6915}$$

$$= 0.7231$$

Worked Example 7.12

A machine is known to produce bolts with a mean diameter of 12 mm and a standard deviation of 0.3 mm. The bolts are too small if the diameter is less than 11.6 mm and considered too large if the diameter exceeds 12.5 mm. Given that the diameters are normally distributed, find the percentage of defective bolts.

It costs 5p to produce a bolt and they are acceptable for use if they are between the limits stated above. If a bolt is too small in diameter it has a scrap value of 1p. If the diameter is too large, a further machine operation is required at an additional cost of 2p to produce a bolt within the acceptable limits. Find the cost per usable bolt.

Solution 7.12

Using the standardisation formula

Standardising 11.6

$$z = \frac{11.6 - 12.0}{0.3} = -1.33$$

Standardising 12.5

$$z = \frac{12.5 - 12.0}{0.3} = 1.67$$

$$\begin{aligned} P(\text{Defective bolt}) &= P(Z < -1.33 \text{ or } Z > 1.67) \\ &= P(Z < -1.33) + P(Z > 1.67) \\ &= 0.0918 + 0.0475 \\ &= 0.1393 \end{aligned}$$

So the proportion of defective bolts is 13.9%.

The second part of the problem necessitates the calculation of an expectation.

P(Bolt is scrapped) $= 0.0918$
P(Bolt is between acceptable limits) $= 0.8607$
P(Bolt has a further machine operation) $= 0.0475$
P(Usable bolt) $= 0.8607 + 0.0475 = 0.9082$

Therefore the average cost per usable bolt

$$= \frac{\text{Expected cost per bolt}}{P(\text{usable bolt})}$$

$$= \frac{4 \times 0.0918 + 5 \times 0.8607 + 7 \times 0.0475}{0.9082}$$

$$= 5.51 \text{ pence}$$

Worked Example 7.13

A toy shop runs a credit card facility for 600 customers. On a particular month, the balances outstanding were as given in Table 7.2.

Table 7.2

Balance	*Frequency*
1-20	30
21-40	90
41-60	400
61-100	70
greater than 100	10

Use the Normal distribution, stating clearly any approximation you make, to calculate the expected frequencies. Comment on the suitability of the results.

Solution 7.13

When tackling a grouped frequency distribution of this type we assume that the variates in each class lie evenly on each side of the class mid-point. We also have to consider an upper bound for the last class interval. In this case we shall choose 200. By using appropriate methods and without using Sheppard's correction factor, the mean is found to be 50.67 and the standard deviation is 20.28.

In order to calculate the expected frequencies we have to standardise the class boundaries. From these figures we can then calculate the corresponding area for each class and subsequently the expected frequency. Table 7.3 gives the results.

Table 7.3

Balance	*Class boundaries*	*z-value*	*Area for each class*	*Expected frequency*
	0.5	-2.47		
1-20			0.0613	37
	20.5	-1.49		
21-40			0.2404	144
	40.5	-0.50		
41-60			0.3759	225
	60.5	0.48		
61-100			0.3087	185
	100.5	2.46		
101-200			0.0069	4
	200.5	>6		

Clearly the results are not a good fit to the actual frequencies owing to the skewness of the given distribution.

Worked Example 7.14

The probability of a certain variety of flower growing from a seedling is 0.4. If 10 seedlings are planted what is the probability of at least 3 growing? If the number of seedlings planted was increased to 100 use a suitable approximation to find the probability of less than 30 plants growing.

Solution 7.14

Using the binomial distribution

P(At least 3 plants growing) = $1 - P(0, 1 \text{ and } 2$ plants growing)

$= 1 - ((0.6)^{10} + 10(0.6)^9\ (0.4) + 45(0.6)^8\ (0.4)^2$

$= 1 - 0.1673$

$= 0.8327$

For the second part we use the normal approximation to the binomial distribution and we standardise 29.5.

$\mu = np = 40$

$\sigma = \sqrt{npq} = 4.9$

Standardising 29.5

$$z = \frac{29.5 - 40}{4.9} = -2.14$$

P(less than 30 plants growing) $= P(Z < -2.14)$
$= 0.0162$

Worked Example 7.15

Blue-blindness, a form of colour blindness, affects 1 in 20 000 people. In a city of some 500 000 people, find the probability of more than 30 people suffering from blue-blindness.

Solution 7.15

The calculations involved with the relevant Poisson distribution would be far too cumbersome for the solution to this problem. We can however use the Normal approximation to the Poisson.

The mean of the Poisson = 500 000/20 000 = 25
The standard deviation = $\sqrt{25} = 5$

Allowing for the continuity correction factor we standardise 30.5.
Standardising 30.5

$$z = \frac{30.5 - 25}{5} = 1.1$$

P(More than 30 suffering from blue-blindness) $= P(Z > 1.1) = 0.1357$

Worked Example 7.16

A baker makes digestive biscuits whose masses are normally distributed with mean 24.0 g and standard deviation 1.9 g. The biscuits are packed by hand into packets of 25. Assuming the biscuits included in each packet are a random sample from the population, what is the distribution of the total mass of biscuits in a packet and what is the probability that it lies between 598 g and 606 g?

Ten packets of biscuits are placed in a box. What is the probability that the total mass of biscuits in the box lies between 6010 g and 6060 g?

A new packer was including 26 biscuits in each packet. What is the probability that a packet selected at random from those containing 25 biscuits would contain a greater mass of biscuits than a packet selected at random from those containing 26 biscuits? (AEB 1984)

Solution 7.16

Let X be the distribution of the combined mass of 25 biscuits. X is normally distributed

$\mu_X = 25 \times 24 = 600$ g

$\sigma_X = \sqrt{25 \times 1.9^2} = 9.5$ g

To find $P(598 < X < 606)$ we standardise 598 and 606

$$z_1 = \frac{598 - 600}{9.5} = -0.21$$

$$z_2 = \frac{606 - 600}{9.5} = 0.63$$

So $P(598 < X < 606) = P(-0.21 < Z < 0.63)$
$= 1 - P(Z < -0.21 \text{ or } Z > 0.63)$
$= 1 - (0.4168 + 0.2643)$
$= 1 - 0.6811$
$= 0.319$

If Y is the distribution of the combined mass of ten packets, Y is also normally distributed.

$\mu_Y = 10 \times 600 = 6000$ g

$\sigma_Y = \sqrt{250 \times 1.9^2} = 30.04$ g

Standardising 6010 and 6060

$$z_3 = \frac{6010 - 6000}{30.04} = 0.33$$

$$z_4 = \frac{6060 - 6000}{30.04} = 2.00$$

So $P(6010 < Y < 6060) = P(0.33 < Z < 2.00)$
$= 0.3707 - 0.02275$
$= 0.348$

Let W represent the distribution of the combined mass of 26 biscuits.

$\mu_W = 26 \times 24 = 624$ g

$\sigma_W = \sqrt{26 \times 1.9^2} = 9.69$ g

The question requires us to find $P(X - W > 0)$

$\mu_{X-W} = 600 - 624 = -24$ g

$\sigma_{X-W} = \sqrt{25 \times 1.9^2 + 26 \times 1.9^2} = 13.57$ g

Standardising 0

$$z = \frac{0 - (-24)}{13.57} = 1.77$$

$P(X - W > 0) = P(Z > 1.77)$

$= 0.0384$

7.7 Exercises

Exercise 7.1

The marks on an English test were standardised. Given that a mark of 52 is equivalent to -1 and a mark of 91 is equivalent to 1.6 standard units, calculate the standard score of someone with a mark of 85.

Exercise 7.2

The acceptable limits of the contact breaker gap in a distributor on a car are 0.7 $\pm$ 0.1 mm. Given that the machine which sets the original gap produces a failure rate of 2.4% too small and 1.2% too large a gap, find the mean and standard deviation at which the machine was operating, assuming a Normal distribution. Find the proportion rejected if the standard deviation remained the same but the mean was set to 0.65 mm.

Exercise 7.3

The mean time of arrival of 500 students at a college is 08.52 a.m. and the standard deviation is 5 minutes. Assuming that the times are normally distributed, find how many students arrive (a) between 08.45 a.m. and 08.55 a.m. (b) after 09.00 a.m.

At a second college, the arrival times of the students follow exactly the same distribution. Given that 15 students are late arriving after 09.00 a.m. estimate the number of students at the college.

Exercise 7.4

A golfer can hit a golf ball a distance of at least 220 metres three times in every five tee shots and a distance of at least 240 metres once in every five tee shots. Given that the distance he clears is normally distributed, find the probability that his drive from the tee ends in a stream 10 metres wide which crosses the fairway, the nearest bank of the stream being 195 metres from the tee.

Exercise 7.5

A survey was carried out over a period of time to record the number of visits a year made to the cinema. The number of visits is shown in the following table:

Number	0	1	2	3	4	5	6	7	8	9	10 or more
Frequency	1	3	7	20	32	30	19	8	2	3	0

Assuming that this sample fits a normal distribution calculate an estimate for the proportion of the population who visit the cinema 6 or more times in a year.

Exercise 7.6

A politician for a local election is informed that the probability that anyone will answer the door when he canvasses is 0.8. Calculate the probability in a row of

10 houses that 7 or less answer the door. By using a suitable approximation, estimate the probability that at least 75 answer the door if he knocks on 100 doors.

Exercise 7.7

A sample group of 4000 people are tested using a new drug. It is known that the probability of an allergic reaction is 0.004. Using a suitable approximation, find the probability that more than 20 people are affected by the reaction.

Exercise 7.8

An industrial control unit contains 12 identical components and a base unit. The components have a mass which is normally distributed with mean 6.8 g and standard deviation 0.8 g. The base unit has a mass which is normally distributed with mean 16.4 g and standard deviation 0.62 g.

(a) Find the probability that the total mass of the complete unit
  (i) exceeds 100 g,
  (ii) is less than 95 g.

(b) The unit costs £2.75 to manufacture and is sold for £6.50 if the mass lies between the limits calculated in part (a); otherwise they are sold for £1 each. Calculate the expected profit on a batch of 1000 units.

Exercise 7.9

Two teams comprising of 4 men and 4 women were chosen by the tiny principality of Sergonia to represent their country in the next Olympics. Given that the heights of men from Sergonia are normally distributed with mean 175 cm and standard deviation 7 cm and the heights of women are also normally distributed with mean 169 cm and standard deviation 5 cm, calculate the probability that:

(i) all four men exceed 173 cm in height;
(ii) two women, chosen at random, differ in height by at least 5 cm;
(iii) a man and a woman chosen at random differ in height by at least 5 cm.

Exercise 7.10

From a large population, a sample of 100 plastic rods is taken and their lengths measured and recorded as follows:

Length (cm)	14.0–	14.2–	14.4–	14.6–	14.8–
Frequency	1	4	11	13	21
Length (cm)	15.0–	15.2–	15.4–	15.6–	15.8–16.0
Frequency	24	12	9	3	2

Calculate the mean length of a rod in the sample and the standard deviation of this length. Assuming that the population is normally distributed with this mean and standard deviation, calculate the percentage of rods with a length between 14.5 cm and 15.5 cm. (UCLES)

Exercise 7.11

The lengths, in centimetres, of manufactured rods have a continuous distribution with probability density function

$f(x) = k(7-x) \qquad 1.9 \leqslant x \leqslant 2.1,$

$f(x) = 0, \qquad$ otherwise.

where k is a constant. A medial rod is defined to be one whose length lies in the range from 1.96 cm to 2.04 cm.

(i) Find the value of k.
(ii) Show that there is a probability of 0.4 that a randomly chosen rod will be a medial rod.
(iii) If four rods are chosen at random find the probability that exactly three of them will be medial rods.
(iv) If 96 rods are chosen at random find an approximate value for the probability that at least 50 of them will be medial rods.
(v) A person requires 20 medial rods for a specific purpose. Find the smallest number of rods the person should order for the expected number of medial rods among those ordered to be at least 20. (W)

Exercise 7.12

The manager of a company producing a large number of bags of flour to a nominal weight of 2 kg has two possible production processes, A and B, at his disposal. In both cases the weights are normally distributed about a mean of 2.05 kg. In A, the standard deviation is 0.025 kg and the profit per bag of flour sold is 5p. In B, the standard deviation is σ kg, which is such that the probability that the weight of a bag of flour will be less than 2 kg is 0.01, but on account of the additional labour costs involved the profit per bag sold is only 4p.

The manager has also to take into account that for each bag of flour sold which weighs less than 2 kg there is a probability of 0.001 that the purchaser will initiate legal proceedings against him, under fair trading laws, which will lead to his company having to pay a fine of £500.

(i) Find, correct to four decimal places, the probability that in process A the weight of a bag of flour is less than 2 kg.
(ii) Find the value of σ, correct to 3 significant figures.
(iii) Determine which process will in the long term produce the greater profit.

(Assume that all bags of flour produced are sold and that the weight of an empty bag is negligible.) (UCLES)

Exercise 7.13

A man travels to work by train and bus. His train is due to arrive at 0845 and the bus he hopes to catch is due to leave the station at 0848. The time of arrival of the train has a Normal distribution with mean at 0844 and standard deviation 2 minutes; the departure time of the bus is independently normally distributed with mean at 0848 and standard deviation 1 minute. Calculate, giving your answers correct to 2 significant figures in (i) and (ii), and to 1 significant figure in (iii), the probabilities that:

(i) the train is late,
(ii) the bus departs before the train arrives,
(iii) in a period of 5 days there are at least 3 days on which the bus departs before the train arrives. (UCLES)

Exercise 7.14

The weights of the contents of cans of fruit are normally distributed with mean 250.2 g and standard deviation 2 g.

(i) Calculate the proportion of the cans that contain less than 250 g.
(ii) Given that 75% of the cans contain at least w g, find the value of w correct to one decimal place.
(iii) Find the probability that the combined contents of four cans will weigh more than 1 kg.
(iv) The weights of the filled cans are distributed with mean 274.5 g and standard deviation 2.5 g. Assuming that the weight of the contents of a can is independent of the weight of the can when empty, determine the mean and the standard deviation of the weights of the empty cans. (W)

Exercise 7.15

Yuk Ping belongs to an athletic club. In javelin-throwing competitions her throws are normally distributed with mean 41.0 m and standard deviation 2.0 m.

(a) What is the probability of her throwing between 40 m and 46 m?
(b) What distance will be exceeded by 60% of her throws?

Gwen belongs to the same club. In competitions 85% of her javelin throws exceed 35 m and 70% exceed 37.5 m. Her throws are normally distributed.

(c) Find the mean and standard deviation of Gwen's throws, each correct to two significant figures.
(d) What is the probability that, in a competition in which each athlete takes a single throw, Yuk Ping will beat Gwen?
(e) The club has to choose one of these two athletes to enter a major competition. In order to qualify for the final rounds it is necessary to achieve a throw of at least 48 m in the preliminary rounds. Which athlete should be chosen and why? (AEB 1990)

*8 Generating Functions

8.1 Probability Generating Functions

We have seen in previous chapters, standard methods for calculating parameters of both discrete and continuous probability distributions. Sometimes, as in the case of the binomial distribution, the derivation of such parameters as the mean and the variance can prove to be quite arduous. We shall now see how the task can be made a little easier if we can define a *probability generating function* for a particular discrete probability distribution.

For a discrete random variable X which can take possible values 0, 1, 2, 3 . . , with associated probabilities a function $G_X(t)$ defined as

$$G_X(t) = \Sigma p_r t^r = p_0 + p_1 t + p_2 t^2 + p_3 t^3 + \ldots$$

where the coefficient of $t^r = P(X = r)$

is called a probability generating function of X.

Using expectation notation, $G_X(t) = E(t^X)$

Worked Example 8.1

A discrete random variable X takes the values of 1, 2, 3 with probabilities 1/6, 1/2, 1/3. Define the probability generating function.

Solution 8.1

$$G_X(t) = \frac{1}{6}t + \frac{1}{2}t^2 + \frac{1}{3}t^3$$

Properties

(i) $G_X(1) = p_1 + p_2 + p_3 + p_4 + \ldots$

$$= \Sigma p_r$$

$$= 1$$

(ii) If X and Y are discrete probability distributions such that

$Y = aX + b$ then $G_Y(t) = E(t^{aX+b})$

$$= t^b E(t^{aX})$$

$$= t^b G_X(t^a)$$

(iii) If X and Y are independent discrete probability distributions

$$G_{X+Y}(t) = E(t^{X+Y})$$

$$= E(t^X t^Y)$$

$$= E(t^X)E(t^Y)$$

$$= G_X(t)G_Y(t)$$

Worked Example 8.2

Two fair dice are thrown. If X represents the score with the first die and Y represents the score with the second die, find the probability generating function of the total score.

Solution 8.2

$$G_X(t) = G_Y(t) = \frac{1}{6}(t + t^2 + t^3 + t^4 + t^5 + t^6)$$

$$\text{So} \quad G_{X+Y}(t) = \left\{\frac{1}{6}(t + t^2 + t^3 + t^4 + t^5 + t^6)\right\}^2$$

Worked Example 8.3

Derive the probability generating function for the binomial distribution.

Solution 8.3

Let X be the discrete random variable denoting the number of successes in n independent trials and let X_i be the random variable such that

$X_i = 1$ if the ith trial is successful

$X_i = 0$ if the ith trial is a failure

If p is the probability of success and q is the probability of failure

then $\quad G_{X_i}(t) = q + pt$

Now $\quad X = X_1 + X_2 + X_3 + \ldots + X_n$

where each X_i is independent

So $G_X(t) = G_{X_1}(t)G_{X_2}(t)G_{X_3}(t) \ldots G_{X_n}(t)$

$\Rightarrow G_X(t) = (q + pt)^n$

8.2 Expectation and Variance using Probability Generating Functions

Now $\quad G_X(t) = p_0 + p_1 t + p_2 t^2 + p_3 t^3 + p_4 t^4 + \ldots$

differentiating with respect to t,

$$G'_X(t) = p_1 + 2p_2 t + 3p_3 t^2 + 4p_4 t^3 + \ldots$$

and putting $t = 1$, $\quad G'_X(1) = p_1 + 2p_2 + 3p_3 + 4p_4 + \ldots$

$$= E(X)$$

To derive an expression for the variance, we need to differentiate again

$$G''_X(t) = 2p_2 + 6p_3 t + 12p_4 t^2 + \ldots$$

and putting $t = 1$, $\quad G''_X(1) = 2p_2 + 6p_3 + 12p_4 + \ldots$

from above $\quad G'_X(1) = p_1 + 2p_2 + 3p_3 + 4p_4 + \ldots$

and adding $\quad G''_X(1) + G'_X(1) = p_1 + 4p_2 + 9p_3 + 16p_4 + \ldots$

$$= E(X^2)$$

Now $\quad \text{Var}(X) = E(X^2) - \{E(X)\}^2$

$$= G''_X(1) + G'_X(1) - \{G'_X(1)\}^2$$

Worked Example 8.4

For the discrete probability distribution X defined in Worked Example 8.1 find the mean and the variance.

Solution 8.4

$$G_X(t) = \frac{1}{6}t + \frac{1}{2}t^2 + \frac{1}{3}t^3$$

$$G'_X(t) = \frac{1}{6} + t + t^2 \Rightarrow G'_X(1) = \frac{13}{6}$$

and $\quad G''_X(t) = 1 + 2t \Rightarrow G''_X(1) = 3$

So $\quad E(X) = G'_X(1) = 13/6$

and $\quad \text{Var}(X) = G''_X(1) + G'_X(1) - \{G'_X(1)\}^2$

$$= 3 + 13/6 - 169/36$$

$$= 17/36$$

8.3 Moment Generating Functions

We shall now define a generating function which can be used for both discrete and continuous probability distributions.

The *moment generating function* of a random variable X is defined as $E(e^{Xt})$.

For a discrete distribution,

$$M_X(t) = \Sigma p_r e^{rt}$$

and for a continuous distribution $f(x)$

$$M_X(t) = \int_{-\infty}^{\infty} e^{xt} f(x)\,dx$$

Worked Example 8.5

Find the moment generating function of the distribution given by $f(x) = 2e^{-2x}$ $(x \geqslant 0)$

Solution 8.5

$$M_X(t) = E(e^{Xt})$$

$$= \int_0^{\infty} e^{xt}\, 2e^{-2x}\,dx$$

$$= \int_0^{\infty} 2\,e^{-(2-t)x}\,dx$$

$$= 2\left[\frac{-e^{-(2-t)x}}{2-t}\right]_0^{\infty}$$

$$= \frac{2}{2-t}$$

Properties

(i) $M_X(0) = 1$

(ii) If X and Y are probability distributions such that $Y = aX + b$

$$\begin{aligned} M_Y(t) &= E(e^{(aX+b)t}) \\ &= e^{bt}E(e^{aXt}) \\ &= e^{bt}M_X(at) \end{aligned}$$

(iii) If X and Y are independent probability distributions

$$\begin{aligned} M_{X+Y}(t) &= E(e^{(X+Y)t}) \\ &= E(e^{Xt}e^{Yt}) \\ &= E(e^{Xt})E(e^{Yt}) \\ &= M_X(t)M_Y(t) \end{aligned}$$

(iv) For a discrete random variable X with a defined probability distribution

$$\begin{aligned} M_X(t) &= E(e^{Xt}) \\ &= E((e^t)^X) \end{aligned}$$

$$\text{So } M_X(t) = G_X(e^t)$$

For the binomial distribution

$$G_X(t) = (q + pt)^n;\ M_X(t) = (q + pe^t)^n$$

Worked Example 8.6

A probability distribution has a moment generating function defined as

$$M_X(t) = \frac{1}{1 - t/\lambda}$$

Find the moment generating function of Y, where $Y = 2X - 3$.

Solution 8.6

$$\begin{aligned} M_Y(t) &= e^{-3t}M_X(2t) \\ &= \frac{e^{-3t}}{1 - 2t/\lambda} \end{aligned}$$

8.4 Expectation and Variance Using Moment Generating Functions

For a discrete random variable X with a defined distribution we proceed in a similar manner to the derivation of the mean and variance using probability generating functions.

$$\begin{aligned} M_X(t) &= G(e^t) \\ &= p_0 + p_1e^t + p_2e^{2t} + p_3e^{3t} + \ldots \quad \ldots \end{aligned}$$

differentiating with respect to t,

$$M'_X(t) = p_1e^t + 2p_2e^{2t} + 3p_3e^{3t} + \ldots$$

and putting $t = 0$, $\quad M'_X(0) = p_1 + 2p_2 + 3p_3 + \ldots$

$$= E(X)$$

Differentiating again,

$$M''_X(t) = p_1 e^t + 4p_2 e^{2t} + 9p_3 e^{3t} + \ldots$$

and putting $t = 0$ $\quad M''_X(0) = p_1 + 4p_2 + 9p_3 + \ldots$

$$= E(X^2)$$

So $\quad \text{Var}(X) = M''_X(0) - \{M'_X(0)\}^2$

For a continuous probability distribution $f(x)$

$$M_X(t) = \int_{-\infty}^{\infty} e^{xt} f(x)\,dx$$

$$= \int_{-\infty}^{\infty} \left(1 + xt + \frac{(xt)^2}{2!} + \frac{(xt)^3}{3!} \ldots\right) f(x)\,dx$$

$$= \int_{-\infty}^{\infty} f(x)\,dx + t\int_{-\infty}^{\infty} xf(x)\,dx + \frac{t^2}{2!}\int_{-\infty}^{\infty} x^2 f(x)\,dx + \frac{t^3}{3!}\int_{-\infty}^{\infty} x^3 f(x)\,dx + \ldots$$

$$= 1 + E(X)t + E(X^2)\frac{t^2}{2!} + E(X^3)\frac{t^3}{3!} + \ldots$$

By differentiation we can verify that the results derived for discrete probability distributions are also true in the continuous case.

A corollary from the last derivation is that $E(X^r)$ is the coefficient of $t^r/r!$.

Worked Example 8.7

Find the mean and variance of the distribution defined in Worked Example 8.5.

Solution 8.7

The moment generating function was found to be

$$M_X(t) = \frac{2}{2-t}$$

$$= 2(2-t)^{-1}$$

$$= (1 - t/2)^{-1}$$

$$= 1 + \frac{t}{2} + \frac{t^2}{4} + \frac{t^3}{8} + \ldots$$

Now $\quad E(X) = \text{coefficient of } t = 1/2$

and $\quad E(X^2) = \text{coefficient of } t^2/2! = 1/2$

So $\quad \text{Var}(X) = \frac{1}{2} - \left(\frac{1}{2}\right)^2 = \frac{1}{4}$

Worked Example 8.8

A discrete random variable X with a defined probability distribution has a probability generating function $G_X(t) = \frac{8}{27}\left(1 + \frac{1t}{2}\right)^3$. Find

(i) $P(X = 2)$, (ii) $E(X)$, (iii) $\text{Var}(X)$

Solution 8.8

$$\text{(i)}\quad G_X(t) = \frac{8}{27} + \frac{4}{9}t + \frac{2}{9}t^2 + \frac{1}{27}t^3$$

$$\text{so}\quad P(X = 2) = \text{coefficient of } t^2 = 2/9$$

$$\text{(ii)}\quad G'_X(t) = \frac{4}{9} + \frac{4}{9}t + \frac{1}{9}t^2$$

$$E(X) = G'_X(1) = 1$$

$$\text{(iii)}\quad G''_X(t) = \frac{4}{9} + \frac{2}{9}t$$

$$G''_X(1) = 2/3$$

$$\text{So}\quad \text{Var}(X) = \frac{2}{3} + 1 - (1)^2 = \frac{2}{3}$$

Worked Example 8.9

A discrete random variable X has a probability density function

$P(X = r) = kr$ for $r = 0, 1, 2, 3, 4$

where k is a constant. Find k and write down the probability generating function. Hence calculate the mean.

Solution 8.9

As $\sum_r p_r = 1$ $\quad k(0 + 1 + 2 + 3 + 4) = 1 \Rightarrow k = 1/10$

The probability generating function

$$G_X(t) = \frac{1}{10}(t + 2t^2 + 3t^3 + 4t^4)$$

$$E(X) = G'_X(1) = (1 + 4 + 9 + 16)/10$$

$$= 3$$

Worked Example 8.10

For the binomial distribution

$$P(X = r) = \binom{n}{r} q^{n-r} p^r$$

Find the mean and the variance.

Solution 8.10

The probability generating function for the defined distribution is

$$G_X(t) = (q + pt)^n$$

So $$G'_X(t) = np(q + pt)^{n-1}$$

$$E(X) = G'_X(1) = np \qquad (q + p = 1)$$

and $$G''_X(t) = n(n-1)p^2(q + pt)^{n-2}$$

$$\Rightarrow G''_X(1) = n(n-1)p^2$$

$$\begin{aligned}\text{Var}(X) &= G''_X(1) + G'_X(1) - \{G'_X(1)\}^2 \\ &= n(n-1)p^2 + np - (np)^2 \\ &= np(1-p) \\ &= npq\end{aligned}$$

Worked Example 8.11

Figure 8.1 shows a maze. A rat is placed at position *. Given that it chooses an exit from a room totally at random, derive the probability generating function for the probability distribution X, where X is the number of rooms visited before reaching the food and hence find the mean number of rooms visited.

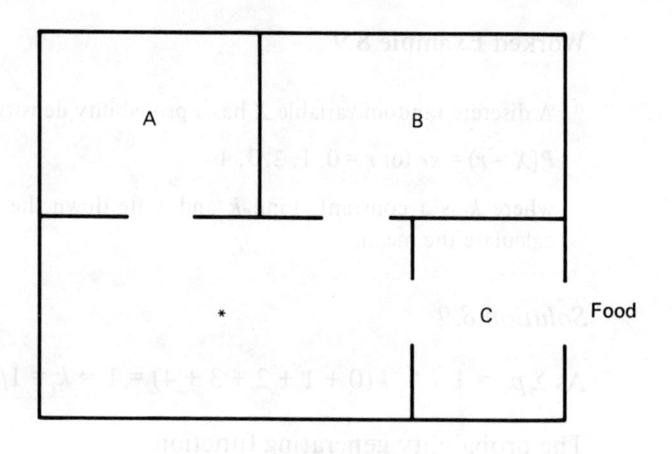

Figure 8.1

Solution 8.11

There are three points to bear in mind:

(i) there are an infinite number of probabilities,
(ii) X can only take odd values,
(iii) to reach the food, the rat must pass through C and therefore we need to consider the probability of entering C and taking the correct exit.

So $P(X = 1) = \frac{1}{3} \cdot \frac{1}{2} = \frac{1}{6}$

(The rat enters C and leaves through door to food.)

$$P(X=3) = \frac{5}{6}\cdot\frac{1}{3}\cdot\frac{1}{2} = \frac{5}{6}\cdot\frac{1}{6}$$

(This is the probability that the rat does not make it to the food by passing through one room (5/6) but on its second move it must be back at * and therefore we combine 5/6 with $P(X=1)$.)

$$P(X=5) = \frac{5}{6}\cdot\frac{5}{6}\cdot\frac{1}{3}\cdot\frac{1}{2} = \left(\frac{5}{6}\right)^2\frac{1}{6}$$

(This is the probability that the rat fails to reach the food by passing through one room or by passing through three rooms ($5/6 \times 5/6$) but on its fourth move it must be back at * and again we combine $(5/6)^2$ with $P(X=1)$.)

Recognising a pattern in the derived probabilities, we can now write down the probability generating function.

$$\begin{aligned}
G_X(t) &= \frac{1}{6}t + \left(\frac{5}{6}\right)\frac{1}{6}t^3 + \left(\frac{5}{6}\right)^2\frac{1}{6}t^5 + \left(\frac{5}{6}\right)^3\frac{1}{6}t^7 \ldots \\
&= \frac{1}{6}t\left(1 + \frac{5}{6}t^2 + \left(\frac{5}{6}\right)^2 t^4 + \left(\frac{5}{6}\right)^3 t^6 \ldots\right) \\
&= \frac{1}{6}t\cdot\frac{1}{1-5t^2/6} \quad \text{(An infinite geometric progression.)} \\
&= \frac{t}{6-5t^2}
\end{aligned}$$

Now $\quad G'_X(t) = (6-5t^2)^{-1} + 10t^2(6-5t^2)^{-2}$

So $E(X) = G'_X(1) = 1 + 10 = 11$

Worked Example 8.12

By using the moment generating function for the Poisson distribution with parameter m, show that the mean is m and the standard deviation is $m^{\frac{1}{2}}$.

Solution 8.12

$$P(X=r) = \frac{e^{-m}m^r}{r!} \qquad (r = 0, 1, 2, \ldots)$$

$$\begin{aligned}
M_X(t) &= E(e^{Xt}) \\
&= e^{-m} + e^{-m}me^t + e^{-m}\frac{m^2}{2!}e^{2t} + e^{-m}\frac{m^3}{3!}e^{3t} + \ldots \\
&= e^{-m}\left\{1 + me^t + \frac{(me^t)^2}{2!} + \frac{(me^t)^3}{3!} + \ldots\right\} \\
&= e^{-m}e^{me^t} \\
&= e^{me^t - m}
\end{aligned}$$

$$M'_X(t) = me^t(e^{me^t - m})$$

So $E(X) = M'_X(0) = m$

and $\quad M''_X(t) = me^t(e^{me^t - m}) + m^2e^{2t}(e^{me^t - m})$

$$M''_X(0) = m + m^2$$

$$\text{Var}(X) = M''_X(0) - \{M'_X(0)\}^2 = m + m^2 - m^2$$

So the standard deviation $= \sqrt{\{\text{Var}(X)\}} = m^{\frac{1}{2}}$

Worked Example 8.13

A continuous random variable X has a probability density function defined as

$f(x) = xe^{-x} \qquad (x \geqslant 0)$

By using the moment generating function show that the mean and variance are both 2.

Solution 8.13

$$\begin{aligned}
M_X(t) &= E(e^{Xt}) \\
&= \int_0^\infty xe^{-x}e^{xt}\,dx \\
&= \int_0^\infty xe^{-x(1-t)}\,dx \\
&= \left[x\int e^{-x(1-t)}dx - \iint e^{-x(1-t)}dx\,\frac{d(x)}{dx}\,dx\right]_0^\infty \\
&= \left[\frac{-xe^{-x(1-t)}}{1-t} + \int \frac{e^{-x(1-t)}}{1-t}\,dx\right]_0^\infty \\
&= \left[\frac{-xe^{-x(1-t)}}{1-t} - \frac{e^{-x(1-t)}}{(1-t)^2}\right]_0^\infty \\
&= \frac{1}{(1-t)^2} \\
&= (1-t)^{-2} \\
&= 1 + 2t + 3t^2 + \ldots \\
E(X) &= \text{coefficient of } t = 2 \\
E(X^2) &= \text{coefficient of } t^2/2! = 6
\end{aligned}$$

So $\quad \text{Var}(X) = E(X^2) - \{E(X)\}^2 = 6 - 4 = 2$

Worked Example 8.14

We write 'X is $B(n, p)$' to mean that a random variable X has the binomial distribution given by

$$P(X = k) = \binom{n}{k}p^k(1-p)^{n-k}, \; 0 \leqslant k \leqslant n$$

Find the moment generating function of X, when X is $B(n, p)$.

Deduce that if X and Y are independent, and X is $B(n, p)$ and Y is $B(m, p)$, then $X + Y$ is $B(n + m, p)$.

Deduce also the distribution of $n - X$, when X is $B(n, p)$. (OLE)

Solution 8.14

$$\begin{aligned}
M_X(t) &= E(e^{Xt}) \\
&= (1-p)^n + \binom{n}{1}(1-p)^{n-1}pe^t + \binom{n}{2}(1-p)^{n-2}p^2e^{2t} + \ldots + p^n e^{nt} \\
&= (1-p)^n + \binom{n}{1}(1-p)^{n-1}(pe^t) + \binom{n}{2}(1-p)^{n-2}(pe^t)^2 + \ldots + (pe^t)^n \\
&= ([1-p] + pe^t)^n
\end{aligned}$$

X is $B(n, p) \Rightarrow M_X(t) = ([1-p] + pe^t)^n$
Y is $B(n, p) \Rightarrow M_Y(t) = ([1-p] + pe^t)^m$

From the properties of moment generating functions, as X and Y are independent,

$$\begin{aligned}
M_{X+Y}(t) &= M_X(t)M_Y(t) \\
&= ([1-p] + pe^t)^n([1-p] - pe^t)^m \\
&= ([1-p] + pe^t)^{n+m}
\end{aligned}$$

which is the moment generating function of the binomial distribution $B(n + m, p)$.

For the last part of the problem let $U = n - X$ and using property (ii) of moment generating functions,

$$\begin{aligned}
M_U(t) &= e^{nt}M_X(-t) \\
&= e^{nt}([1-p] + pe^{-t})^n \\
&= ([1-p]e^t + p)^n
\end{aligned}$$

which is the moment generating function of the binomial distribution $B(n, 1 - p)$.

8.5 Exercises

Exercise 8.1

A discrete random variable X with a defined probability distribution has a probability generating function $G_X(t) = a + bt + ct^2$. If the mean is 5/4 and the variance is 11/16, find the values of a, b and c.

Exercise 8.2

A game consists of rolling two dice, one six-sided and the other four-sided, and adding the scores together. If both dice are fair and the first is numbered 1 to 6 and the second is numbered 1 to 4, show that the probability generating function of Z where Z is the sum of the scores on the two dice is $\dfrac{t^2}{24}\dfrac{(1-t^6)(1-t^4)}{(1-t)^2}$

Exercise 8.3

Two duellists take alternate shots at each other until either one of them scores a 'hit' on the other. If the probability that the first duellist scores a 'hit' is 1/5 and the probability that the second scores a 'hit' is 1/3, find the expected number of shots before a hit.

Exercise 8.4

The variable X has a Poisson distribution with mean λ. Write down the value of $P(X = r)$. Deduce that the probability generating function is given by $G(t) = e^{\lambda(t-1)}$.

The variable Y can take only the values 1, 2, 3, . . ., and is such that $P(Y = r) = kP(X = r)$ where k is a constant, $r > 0$.

Find the value of k.

Show that the probability generating function of Y is $\dfrac{e^{\lambda t} - 1}{e^{\lambda} - 1}$.

(UCLES)

Exercise 8.5

An unbiased coin is tossed n times. If a head appears, +1 is scored; if a tail appears, -1 is scored. The random variable X is the total score obtained. Prove that the probability generating function of X is $\left(\dfrac{z}{2} + \dfrac{1}{2z}\right)^n$.

Hence or otherwise, find

(i) $E(X)$,
(ii) $\text{var}(X)$.

Another unbiased coin is tossed n times. If a head appears, +2 is scored; if a tail appears, -1 is scored. The random variable Y is the total score obtained in these n tosses. Find $\text{var}(X + Y)$. (UCLES)

Exercise 8.6

When a die is thrown, event A occurs with probability p. The random variable N is defined by:

'the event A occurs for the first time on the Nth throw'.

Write down an expression for $P(N = n)$.

Show that $G(t)$, the probability generating function for N, is given by $G(t) = \dfrac{pt}{1 - qt}$, where $p + q = 1$.

Deduce, in terms of p, the value of $E(N)$ and show that $\text{var}(N) = q/p^2$.

An unbiased die is thrown repeatedly until a 5 and 6 have been obtained. The random variable M denotes the number of throws required. (Thus for the sequence of results $\underline{6}$, 3, 1, 3, 6, 6, $\underline{5}$ the value of M is 7.) Show that $E(M) = 9$ and calculate $\text{var}(M)$. (UCLES)

Exercise 8.7

X is uniformly distributed in the interval $-2 \leqslant x \leqslant 2$. Find the moment generating function of X.

Exercise 8.8

There are two outcomes to an event. The probability of success is p. The experiment is continued until a success occurs and X is the distribution of the number of trials up to (but not including) a success. By using a moment generating function, show that $E(X) = \frac{1-p}{p}$.

Exercise 8.9

The continuous random variable X has a probability density function given by

$$f(x) = \begin{cases} 0 & \text{for } x < a \\ \lambda e^{-\lambda(x-a)} & \text{for } x \geqslant a \end{cases}$$

where a and λ are positive constants. Show that the moment generating function of X is given by

$$M(t) = \frac{\lambda e^{at}}{\lambda - t}$$

By obtaining $M'(t)$ and $M''(t)$, or otherwise, find $E(X)$ and $\mathrm{Var}(X)$. (UCLES)

Exercise 8.10

A random variable X is said to have a geometric distribution with parameter α if $P(X = k) = (1 - \alpha)\alpha^{k-1}$ $(k = 1, 2, 3, \ldots)$, where $0 < \alpha < 1$. For such a random variable, find

(i) the probability that $X > k$;
(ii) its moment generating function.

Independent random variables U and V have geometric distributions with parameters β and γ respectively, and $W = \min(U, V)$ is the minimum of U and V. Write down an expression for the probability that $W = k$ in terms of the probabilities that U and V take certain values. Hence find the distribution of W, and deduce its moment generating function. (OLE)

Exercise 8.11

Show that the moment generating function (MGF) of a random variable X having a standard Normal distribution is $\exp(t^2/2)$.

Deduce the value of $E(X^4)$, and also the MGF of Y, where Y has a Normal distribution with mean 0 and variance σ^2.

Y_1, Y_2, Y_3 and Y_4 are independent random variables, each having the same distribution as Y, and k is a constant. By finding the MGF of

$Z = k(Y_1 + Y_2 + Y_3 + Y_4)$, deduce a value of k such that Z is standard normal.

(OLE)

9 Sampling and Statistical Estimation

9.1 Sampling

An arms manufacturer wishes to test the reliability and accuracy of his ammunition. To do this he decides under a controlled experiment, to fire at a target. Clearly it would be pointless to fire all his ammunition at the target as he would have none left to sell. He should instead choose a representative sample of ammunition from the total population of ammunition. (Here 'population' means the manufacturer's total supply of ammunition.) Such a sample could be a *random sample* whereby every item in the population has an equal chance of being selected and each selected item is independent of other items selected.

Providing that the sample is fully representative of the population, the manufacturer can draw conclusions about the distribution of the population.

Sampling is necessary therefore when:

(i) the testing process is destructive as in the example above;
(ii) the population is too large to be completely tested on economical grounds;
(iii) it is impossible to define the population.

From the data collected in a sample we can compute numerical quantities such as the sample mean or sample variance. These can then be used to estimate the corresponding population mean and variance. Such estimates of population parameters are called *point estimates*. If we repeat the sampling procedure from the population and record the point estimates from each sample, by the nature of random sampling these estimates will vary and a distribution will be formed. It is the nature of such distributions which will now concern us.

9.2 Estimators

We shall define a *statistic* as a function of a random sample $X_1, X_2, X_3, \ldots, X_n$, taken from a population.

So for example,

the sample mean $$\bar{X} = \frac{X_1 + X_2 + \ldots + X_n}{n}$$

and the sample variance $$S^2 = \sum \frac{(X - \bar{X})^2}{n}$$

We shall use the equivalent lower case letters to represent the point estimate determined by the numerical values in a particular sample or group of samples.

The statistic is known as an *estimator* of the corresponding population parameter. So $\overline{X}$ is an estimator for the population mean μ and S^2 is an estimator for the population variance, σ^2.

9.3 Distribution of a Statistic

Any statistic is a random variable and therefore has a defined distribution. The standard deviation of a sampling distribution of a statistic is called the *standard error*.

Worked Example 9.1

Find the mean and variance of the distribution $\overline{X}$ defined above where each X is from a distribution with mean μ and variance σ^2.

Solution 9.1

$$\begin{aligned}
E(\overline{X}) &= E\left(\frac{X_1 + X_2 + \ldots + X_n}{n}\right) \\
&= \frac{1}{n}\left\{E(X_1 + X_2 + \ldots + X_n)\right\} \\
&= \frac{1}{n}\left\{E(X_1) + E(X_2) + \ldots + E(X_n)\right\} \\
&= \frac{1}{n}\left\{\mu + \mu + \ldots + \mu\right\} \\
&= \frac{1}{n}.\, n\mu = \mu
\end{aligned}$$

$$\begin{aligned}
\text{Var}(\overline{X}) &= \text{Var}\left(\frac{X_1 + X_2 + \ldots + X_n}{n}\right) \\
&= \frac{1}{n^2}\left\{\text{Var}(X_1 + X_2 + \ldots + X_n)\right\} \\
&= \frac{1}{n^2}\left\{\text{Var}(X_1) + \text{Var}(X_2) \ldots + \text{Var}(X_n)\right\} \\
&= \frac{1}{n^2}\left\{\sigma^2 + \sigma^2 + \ldots + \sigma^2\right\} \\
&= \frac{1}{n^2} n\sigma^2 = \frac{\sigma^2}{n}
\end{aligned}$$

The standard deviation of $\overline{X}$, is therefore $\sigma/\sqrt{n}$ and is the standard error of the mean.

9.4 Central Limit Theorem

If X is a random variable with mean μ and variance σ^2, then the distribution of $\overline{X}$ approximates to a Normal distribution with mean μ and variance σ^2/n, as $n \to \infty$.

Note: (i) The approximation is good for all distributions of X when $n > 30$.
(ii) If X is normally distributed, then $\overline{X}$ is normally distributed for all values of $n > 0$.

This theorem is fundamental to the techniques used in statistical testing and the proof requires rather advanced probability theory.

Worked Example 9.2

Find $E(S^2)$.

Solution 9.2

$$E(S^2) = E\left(\sum\frac{(X-\overline{X})^2}{n}\right)$$

$$= \frac{1}{n}E(\Sigma(X^2 - 2X\overline{X} + \overline{X}^2))$$

$$= \frac{1}{n}E(\Sigma X^2 - 2\overline{X}\Sigma X + n\overline{X}^2)$$

$$= \frac{1}{n}E(\Sigma X^2 - 2n\overline{X}^2 + n\overline{X}^2)$$

$$= \frac{1}{n}E(\Sigma X^2 - n\overline{X}^2)$$

$$= \frac{1}{n}\left\{\Sigma E(X^2) - nE(\overline{X}^2)\right\}$$

$$= \frac{1}{n}\left\{n(\sigma^2 + \mu^2) - n\left(\frac{\sigma^2}{n} + \mu^2\right)\right\}$$

$$= \sigma^2 + \mu^2 - \frac{\sigma^2}{n} - \mu^2$$

$$= \frac{(n-1)}{n}\sigma^2$$

9.5 Unbiased Estimators

An estimator is said to be *unbiased* if the expectation of the estimator is equal to the corresponding population parameter.

Clearly $\overline{X}$ is an unbiased estimator of μ (see Worked Example 9.1) but from Worked Example 9.2, $E(S^2) \neq \sigma^2$

However $E\left(\frac{nS^2}{n-1}\right) = \frac{n}{n-1}E(S^2) = \frac{n}{n-1}\frac{(n-1)}{n}\sigma^2$

$$= \sigma^2$$

so by defining $\hat{S}^2 = \sum\frac{(X-\overline{X})^2}{n-1}$

$\hat{S}^2$ is an unbiased estimator of σ^2. That is $E(\hat{S}^2) = \sigma^2$.

In practice for large n, if we do not know σ^2 and we have to make an estimate from the sample data, the error involved by dividing by n rather than $n-1$ is quite small. However it is important to maintain the correct procedure.

The *best* or *most efficient* estimator of a population parameter is one which gives the smallest possible variance.

An estimator is said to be *consistent* if the variance of the estimator tends to zero as $n \to \infty$.

Worked Example 9.3

A distribution has a known μ and variance σ^2. If two independent observations X_1 and X_2 are made, find the values of a and b such that $aX_1 + bX_2$ is an unbiased and efficient estimator of μ.

Solution 9.3

$$\begin{aligned} E(aX_1 + bX_2) &= aE(X_1) + bE(X_2) \\ &= a\mu + b\mu \\ &= (a+b)\mu \end{aligned}$$

For an unbiased estimator

$$\begin{aligned} E(aX_1 + bX_2) &= \mu \\ \Rightarrow \quad a + b &= 1 \quad \ldots\ldots\ldots\ldots\ldots\ldots (i) \end{aligned}$$

For an efficient estimator we require

$\text{Var}(aX_1 + bX_2)$ to be a minimum

$$\begin{aligned} \text{Now} \quad \text{Var}(aX_1 + bX_2) &= \text{Var}(aX_1) + \text{Var}(bX_2) \leftarrow \boxed{X_1 \text{ and } X_2 \text{ independent}} \\ &= a^2\text{Var}(X_1) + b^2\text{Var}(X_2) \\ &= a^2\sigma^2 + b^2\sigma^2 \\ &= (a^2 + b^2)\sigma^2 \end{aligned}$$

From (i) $a = 1 - b$

$$\begin{aligned} \text{So} \quad a^2 + b^2 &= (1-b)^2 + b^2 \\ &= 2b^2 - 2b + 1 \end{aligned}$$

For a minimum, we require

$$\frac{d}{db}(2b^2 - 2b + 1) = 0$$

$$\begin{aligned} \Rightarrow 4b - 2 &= 0 \\ b &= 1/2 \end{aligned}$$

and substituting into (i) $a = 1/2$

In conclusion, $(X_1 + X_2)/2$ is an unbiased and efficient estimator for μ.

9.6 Interval Estimates and Confidence

Point estimates, by their very nature, give a singular numerical approximation to a population parameter. It is more preferable to use *interval estimates* in which an indication of the precision or accuracy of an estimate can be given.

For the Normal distribution, 95% of the distribution lies within 1.96σ of the mean. This means that if a random value x is chosen from the distribution then $P(\mu - 1.96\sigma < x < \mu + 1.96\sigma) = 0.95$.

Conversely if a random value is chosen from the distribution and we add $\pm 1.96\sigma$ to this value, we can be 95% *confident* that the population mean will lie between the limits $x \pm 1.96\sigma$ and the interval from $x - 1.96\sigma$ to $x + 1.96\sigma$ is said to be the 95% *confidence interval* for the population mean. We shall denote this interval as $(x - 1.96\sigma, x + 1.96\sigma)$.

Worked Example 9.4

A random variable X is known to be normally distributed with a standard deviation of 5. A random observation from this distribution is 12.4. Find the interval such that we can be 99% confident that the population mean lies within this interval.

Solution 9.4

From the Normal distribution tables 99% of random observations lie between $\pm 2.58\sigma$ of the mean.

i.e. $P(\mu - 2.58 . 5 < x < \mu + 2.58 . 5) = 0.99$

So conversely, μ lies in the interval

$$12.4 - 12.9 < \mu < 12.4 + 12.9$$
$$-0.5 < \mu < 25.3$$

with 99% confidence.

-0.5 and 25.3 are the 99% confidence limits for μ and the range of values defined by 12.4 ± 12.9 is the 99% confidence interval for the population mean.

This interval is quite large and not very helpful but it can be reduced in size by taking a random sample of n observations from the population, and by the Central Limit Theorem, the distribution of the mean of this sample will be normally distributed but with a variance of σ^2/n.

Worked Example 9.5

A random variable X is distributed as in Worked Example 9.4. 16 observations from this distribution give a sample mean of 12.4. Find the 99% confidence interval for the population mean.

Solution 9.5

From the Central Limit Theorem, the sample mean $\bar{x}$ is from a Normal distribution with mean μ and standard deviation (standard error)

$$\sigma/\sqrt{n} = 5/4$$

So 99% confidence limits for $\mu = \bar{x} \pm 2.58 \dfrac{\sigma}{\sqrt{n}}$

$$= 12.4 \pm 2.58 . 1.25$$
$$= 12.4 \pm 3.225$$

The 99% confidence interval is (9.175, 15.625).

From this example we should see that to reduce the size of the confidence interval we can:

(i) increase the sample size,
(ii) decrease the level of confidence.

If the population variance is unknown it has to be estimated from the sample data. The unbiased estimate of the population variance is $\hat{s}^2 = \Sigma(x - \overline{x})^2/(n - 1)$ and providing $n > 30$, the sample mean $\overline{x}$ is from a distribution which is approximately Normal with mean μ and standard error $\hat{s}/\sqrt{n}$. If $n \leqslant 30$ and σ^2 is unknown, then providing that we can assume that the population is normally distributed, the small sampling theory of Chapter 11 will apply.

Worked Example 9.6

100 independent observations of a random variable X give $\Sigma x = 58$ and $\Sigma x^2 = 425$. Find the 95% confidence interval for the mean of the distribution.

Solution 9.6

Sample mean $\overline{x} = 58/100 = 0.58$

$$\text{Sample variance } s^2 = \frac{\Sigma(x - \overline{x})^2}{100} = \frac{\Sigma x^2}{100} - \overline{x}^2$$

$$= \frac{425}{100} - (0.58)^2$$

$$= 4.25 - 0.34$$

$$= 3.91$$

But the unbiased estimate of the population variance $\hat{s}^2$

$$= \frac{\Sigma(x - \overline{x})^2}{n - 1}$$

$$= \frac{n}{n - 1} \cdot \frac{\Sigma(x - \overline{x})^2}{n}$$

$$= \left(\frac{n}{n - 1}\right) \times \text{sample variance}$$

$$\text{So the unbiased estimate } \hat{s}^2 = \frac{100}{99} \cdot 3.91$$

$$= 3.95$$

$$\text{The 95\% confidence limits for } \mu = 0.58 \pm 1.96 \sqrt{\frac{3.95}{100}}$$

$$= 0.58 \pm 0.39$$

The 95% confidence interval is (0.19, 0.97).

9.7 Sampling Distribution of Proportions

If we define π as the proportion of a population which possesses a certain property or *attribute*, then the distribution of P, the proportion of items in a random sample of size n possessing the property will have a mean of π and variance $\dfrac{\pi(1 - \pi)}{n}$.

To verify this we must consider the distribution of the number of items in a random sample of size n taken from the population possessing the property. Denoting this distribution by R, then $P = R/n$. As R is binomially distributed with a mean of $n\pi$ and variance of $n\pi(1 - \pi)$

then $$E(P) = E\left(\frac{R}{n}\right) = \frac{1}{n}E(R) = \frac{n\pi}{n}$$
$$= \pi$$

and $$\text{Var}(P) = \text{Var}\left(\frac{R}{n}\right) = \frac{1}{n^2}\text{Var}(R) = \frac{1}{n^2}\left(n\pi(1 - \pi)\right)$$
$$= \frac{\pi(1 - \pi)}{n}$$

In order to calculate a confidence interval for a population parameter π, we have to estimate the standard error using the point estimate, p. So if π is unknown, the standard error of a proportion = $\sqrt{\frac{p(1 - p)}{n}}$ where p is the point estimate of π.

Worked Example 9.7

Of a random sample of 100 train commuters, 60 said that they purchased a paper daily to read on the train. Find the 95% confidence interval for the proportion of train commuters who purchase a paper daily to read on the train.

Solution 9.7

Sample proportion $p = 0.6$

Standard error of the proportion = $\sqrt{\frac{0.6 \cdot 0.4}{100}}$

So the 95% confidence limits $= 0.6 \pm 1.96\sqrt{\frac{0.6 \,.\, 0.4}{100}}$

$= 0.6 \pm 0.096$

The 95% confidence interval is (0.504, 0.696).

Note that the accuracy of the estimate for π is dependent on the sample size and not on the size of the population.

In Chapter 7, when applying the Normal distribution as an approximation to the binomial, we used a continuity correction factor and strictly speaking the same ought to apply here. The sample proportion p is assumed to come from a continuous distribution and therefore applying a continuity correction factor, p should be in the range $\frac{60 \pm 0.5}{100}$. In general, the confidence interval for a proportion should be increased by $+1/2n$ but for relatively large $n\,(n > 50)$ it is usually ignored as the error is small.

9.8 Other Sampling Distributions

(a) Difference between Two Means

X and Y are two populations distributed with means μ_X and μ_Y and variances σ_X^2 and σ_Y^2. The sampling distribution X of all samples of size n_X is distributed with mean μ and variance σ_X^2/n_X. Likewise the sampling distribution of Y of samples of size n_Y is distributed with mean μ_Y and variance σ_Y^2/n_Y.

The sampling distribution of the difference between the sample means, $\overline{X} - \overline{Y}$ is distributed with mean $\mu_X - \mu_Y$ and variance $\dfrac{\sigma_X^2}{n_X} + \dfrac{\sigma_Y^2}{n_Y}$. The distribution $\overline{X} - \overline{Y}$ is approximately normally distributed provided either that

(i) both X and Y are normally distributed,

or (ii) both n_X and n_Y are greater than 30.

If σ_X^2 and σ_Y^2 are unknown and providing n_X and n_Y are both large (>100) then we can use $\hat{s}_X^2$ and $\hat{s}_Y^2$, the unbiased estimates of the population variances, in place of σ_X^2 and σ_Y^2 respectively.

For smaller sample sizes or where it can be considered that the population variances are equal the best estimate of the population variance is

$$\frac{\hat{s}^2}{n_X} + \frac{\hat{s}^2}{n_Y} = \hat{s}^2\left(\frac{1}{n_X} + \frac{1}{n_Y}\right)$$

where $$\hat{s}^2 = \frac{n_X s_X^2 + n_Y s_Y^2}{n_X + n_Y - 2}$$

$\hat{s}^2$ is a *pooled* estimate of the population variance.

Worked Example 9.8

An examination paper consists of two sections, A and B. The maximum mark on each section is 50. The mean and standard deviations for all candidates in both sections is as follows:

	Mean mark	*Standard deviation*
Section A	36	5.5
Section B	42	4.5

If random samples of 100 candidates' marks from section A and 50 candidates' marks from section B are taken, find the probability that the mean mark obtained from the sample from section A will be no less than 5 marks less than the sample mean from section B.

Solution 9.8

The sampling distribution for the differences of the means is $\overline{A} - \overline{B}$.

mean $= \mu_A - \mu_B = 36 - 42$

variance $= \dfrac{\sigma_A^2}{n_A} + \dfrac{\sigma_B^2}{n_B} = \dfrac{5.5^2}{100} + \dfrac{4.5^2}{50}$

and $\overline{A} - \overline{B}$ is normally distributed.

We require $P(\bar{A} - \bar{B} > -5)$ so standardising -5,

$$z = \frac{-5-(-6)}{\sqrt{\frac{5.5^2}{100} + \frac{4.5^2}{50}}} = \frac{1}{0.841}$$

$$= 1.19$$

So $P(\bar{A} - \bar{B} > -5) = P(Z > 1.19) = 0.117$

(b) Difference between Two Proportions

In the first instance we consider the difference between two proportions from two distinct populations. If P_1 and P_2 are the respective sampling distributions of proportions with sample sizes n_1 and n_2, then P_1 and P_2 are both approximately normally distributed for large n. If π_1 and π_2 (the population proportions) are unknown, p_1 and p_2 (the sample proportions) are used as the best estimates for the population parameters and the mean and variances of the distributions P_1 and P_2 are:

$$p_1, \ \frac{p_1(1-p_1)}{n_1} \text{ and } p_2, \ \frac{p_2(1-p_2)}{n_2}$$

As these distributions are independent, the distribution of $P_1 - P_2$ is approximately Normal with mean $p_1 - p_2$ and variance $\dfrac{p_1(1-p_1)}{n_1} + \dfrac{p_2(1-p_2)}{n_2}$.

Worked Example 9.9

In a sample of 1000 people interviewed in Birmingham, 650 had a holiday abroad in 1985. In Southampton a sample of 600 showed that 450 had a holiday abroad in 1985. Estimate a 95% confidence interval for the difference between the two proportions.

Solution 9.9

Difference in sample proportions $= p_1 - p_2 = 0.65 - 0.75$
$= -0.1$

Standard error of the difference of the two proportions

$$= \sqrt{\left\{\frac{p_1(1-p_1)}{n_1} + \frac{p_2(1-p_2)}{n_2}\right\}}$$

$$= \sqrt{\left\{\frac{0.65 \cdot 0.35}{1000} + \frac{0.75 \cdot 0.25}{600}\right\}}$$

$= 0.0232$

So 95% confidence limits are $= (p_1 - p_2) \pm 1.96 \times$ standard error
$= -0.1 \pm 0.0455$

The 95% confidence interval is $(-0.1455, -0.0545)$.

Note: Some questions might refer to a 2σ rule when calculating a confidence interval. This is quite simply adding $\pm$ 2 standard errors to the point estimate of the population parameter. The confidence interval is then slightly larger than the 95% interval. It is often used to make the arithmetic slightly easier.

If there is reason to believe that the population proportions are the same, i.e. $\pi_1 = \pi_2$, then a pooled estimate of the population proportion is made as follows:
In order to find p_1 the number x_1 in the first sample with the desired attribute is divided by n_1, and for p_2 the number x_2 in the second sample with the desired attribute is divided by n_2.

$$\text{Pooled estimate } p = \frac{x_1 + x_2}{n_1 + n_2} = \frac{n_1 p_1 + n_2 p_2}{n_1 + n_2}$$

This estimate then replaces p_1 and p_2 in the expression for a confidence interval.

(c) Distribution of the Mean for Finite Populations

If we sample from a finite population without replacement then the probability of any item being chosen is dependent on what has been chosen previously. In this instance

$$E(\bar{X}) = \mu$$

$$\text{and } \operatorname{Var}(\bar{X}) = \frac{\sigma^2}{n} \times \frac{N - n}{N - 1}$$

where μ are σ are the population parameters, N is the population size and n is the sample size.

(d) Distribution of the Standard Deviation and the Variance

If the population is known to be normally distributed, the standard errors of the two distributions are:

Distribution	Standard Error
Standard Deviation	$\sigma/\sqrt{2n}$
Variance	$\sigma^2\sqrt{\frac{2}{n}}$

Worked Example 9.10

A random variable X is normally distributed with a mean of 12 and a standard deviation of 3. If a random sample of 36 is chosen and found to have a mean $\bar{X}$, find $P(\bar{X} > 13)$. Find also the sample size such that $P(\bar{X} > 12.5) = 0.05$.

Solution 9.10

First part: $\bar{X}$ is normally distributed with mean 12 and standard deviation $3/\sqrt{36} = 1/2$.
Standardising 13

$$z = \frac{13 - 12}{1/2} = 2$$

$$\text{So} \quad P(X > 13) = P(Z > 2) = 0.02275$$

Second part: from the Normal distribution tables, the standard score z such that $P(Z > z) = 0.05$ is 1.645.

Using the standardisation formula

$$1.645 = \frac{12.5 - 12}{3/\sqrt{n}}$$

$$1.645 \cdot \frac{3}{\sqrt{n}} = 0.5$$

$$\frac{1.645 \cdot 3}{0.5} = \sqrt{n}$$

$$\Rightarrow \quad n = (9.87)^2$$

$$\Rightarrow \quad n = 97.42$$

As n has to take an integer value, the required sample size is 98.

Worked Example 9.11

A sample of 50 is drawn from a population which is known to follow a Poisson distribution with mean 2. Find the probability that the sample mean will be greater than 2.5.

Solution 9.11

For a Poisson distribution with mean 2, the standard deviation is $\sqrt{2}$. Samples of size 50 are normally distributed with mean 2 and standard deviation $\sqrt{2/50} = 0.2$.

Standardising 2.5

$$z = \frac{2.5 - 2}{0.2} = 2.5$$

If $\overline{X}$ represents the sampling distribution of the mean,

$$P(\overline{X} > 2.5) = P(Z > 2.5)$$

$$= 0.0062$$

Worked Example 9.12

A random sample of 50 items has a mean of 27.2 and the sum of the squared deviations from this figure is 290. Calculate 95% confidence limits for the population mean.

Solution 9.12

The 'sum of the squared deviations from the mean'

$$= \Sigma(x - \overline{x})^2$$

So the sample variance $s^2 = \dfrac{\Sigma(x - \overline{x})^2}{n} = \dfrac{290}{50}$

$$= 5.8$$

The unbiased estimator of the population variance

$$\hat{s}^2 = \frac{50 \cdot 5.8}{49}$$

$$= 5.92$$

The standard deviation of the sampling distribution of the mean (standard error)

$$= \hat{s}/\sqrt{n} = \sqrt{\frac{5.92}{50}}$$

$$= 0.344$$

The 95% confidence limits $= 27.2 \pm 1.96 . 0.344$

$= 27.2 \pm 0.674$

So to 3 significant figures the 95% confidence limits are 26.5 and 27.9.

Worked Example 9.13

Two independent samples of sizes n_1 and n_2 are taken from the same population. If the sample variances are s_1^2 and s_2^2, show that the pooled variance $= \dfrac{n_1 s_1^2 + n_2 s_2^2}{n_1 + n_2 - 2}$ is an unbiased estimate of the population variance.

Solution 9.13

We have already shown that $E\left(\frac{ns^2}{n-1}\right) = \sigma^2$ where s^2 is the sample variance.

$$\Rightarrow E(ns^2) = (n-1)\sigma^2$$

So $\quad E(n_1 s_1^2) = (n_1 - 1)\sigma^2 \quad$ (i)

and $\quad E(n_2 s_2^2) = (n_2 - 1)\sigma^2 \quad$ (ii)

Adding (i) and (ii) $E(n_1 s_1^2) + E(n_2 s_2^2) = (n_1 - 1)\sigma^2 + (n_2 - 1)\sigma^2$

$$= (n_1 + n_2 - 2)\sigma^2$$

So $\quad E(n_1 s_1^2 + n_2 s_2^2) = (n_1 + n_2 - 2)\sigma^2$

$$\Rightarrow E\left(\frac{n_1 s_1^2 + n_2 s_2^2}{n_1 + n_2 - 2}\right) = \sigma^2$$

Two other useful forms of the pooled variance are

(i) $\dfrac{\Sigma(x_i - \overline{x}_1)^2 + \Sigma(x_j - \overline{x}_2)^2}{n_1 + n_2 - 2}$ where $x_i, x_j, \overline{x}_1, \overline{x}_2$ are sample values and sample means

(ii) $\dfrac{(n_1 - 1)\hat{s}_1^2 + (n_2 - 1)\hat{s}_2^2}{n_1 + n_2 - 2}$ where $\hat{s}_1^2$ and $\hat{s}_2^2$ are the unbiased estimates of the population variance σ^2.

Worked Example 9.14

100 flights by a particular plane to a certain destination gave a mean flight time of 5.93 hours with a standard deviation of 0.3 hours. The flight times of 120 journeys to the same destination by a different type of aircraft gave a mean flight time of 6.02 hours with a standard deviation of 0.4 hours. Stating any assumptions that you make, find a 95% confidence interval for the difference in the mean flying times of the two types of aircraft.

Solution 9.14

The difference between the two standard deviations is small so we shall make the assumption that the two population variances are the same.

The best estimate for the population variance therefore is

$$\hat{s}^2 = \frac{n_1 s_1^2 + n_2 s_2^2}{n_1 + n_2 - 2}$$

$$= \frac{100\,.(.3)^2 + 120\,.(.4)^2}{218}$$

$$= 0.129$$

The standard error of the distribution of the difference between the means

$$= \hat{s}\sqrt{\frac{1}{n_1} + \frac{1}{n_2}}$$

$$= 0.36\sqrt{\frac{1}{100} + \frac{1}{120}}$$

$$= 0.36\,.\,0.135$$

$$= 0.049$$

So the 95% confidence limits for the difference between the means

$$= (5.93 - 6.02) \pm 1.96\,.\,0.049$$

$$= -0.09 \pm 0.095 \text{ hours}$$

The 95% confidence interval is $(-0.185, 0.005)$.

Worked Example 9.15

A point whose coordinates are (X, Y) with respect to rectangular axes is chosen at random where $0 < X < 1$ and $0 < Y < 1$. What is the probability that the point lies inside the circle whose equation is $x^2 + y^2 = 1$?

In a computer simulation 1000 such points were generated and 784 of them lay inside the circle. Obtain an estimate for π and give an approximate 90% confidence interval for your estimate. Show that about 290 000 points need to be selected in order to be 90% certain of obtaining a value of π which will be in error by less than 0.005. (SUJB)

Solution 9.15

First part: P(Point lies inside the circle $x^2 + y^2 = 1$ $(0 < x < 1$ and $0 < y < 1)$

$$= \frac{\text{area of quarter circle shaded}}{\text{area of square}} \quad \text{(see Fig. 9.1)}$$

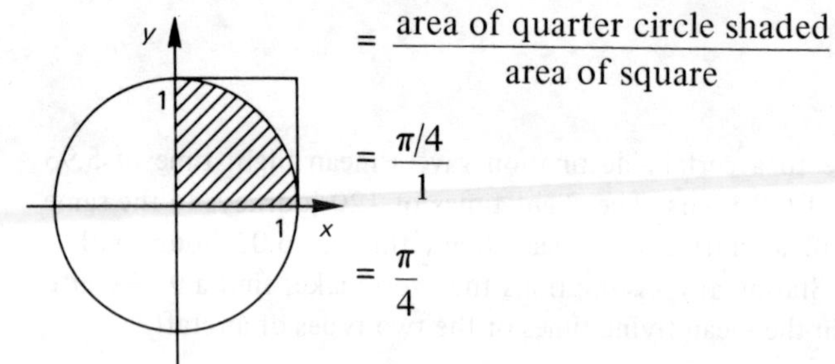

$$= \frac{\pi/4}{1}$$

$$= \frac{\pi}{4}$$

Figure 9.1

Second part:

Point estimate for $\pi/4 = \frac{784}{1000} = 0.784$

So the 90% confidence limits for $\pi/4$

$$= 0.784 \pm 1.645 \sqrt{\frac{0.784 \,.\, 0.216}{1000}}$$

$$= 0.784 \pm 0.021$$

So by multiplying throughout by 4 we find that the 90% confidence limits for π

$= 3.136 \pm 0.084$

The 90% confidence interval for π is therefore (3.052, 3.220).

In order to be in error by less than 0.005 with a 90% certainty

$$4 \,.\, 1.645 \sqrt{\frac{0.784 \,.\, 0.216}{n}} < 0.005$$

$$\frac{16 \,.\, (1.645)^2 \,.\, 0.784 \,.\, 0.216}{(0.005)^2} < n$$

$$293300 < n$$

Which is what was required.

Worked Example 9.16

Paul has a bag containing 8 discs, each one having a number on it. Four of the discs are blue and numbered 2, 3, 3, 4 and the other four are red and numbered 6, 8, 9, 13.

(a) Calculate the mean and variance of this population of eight numbers.

(b) Paul then invites Claire, who does not have any information about the numbers on the discs, to estimate the mean of the numbers. Claire takes a random sample of 4 discs, without replacement. Calculate the mean and variance of the sample mean.

The blue discs are then placed in one bag and the red ones in a separate bag.

(c) Calculate the mean and variance of the numbers in each bag.

(d) Paul then invites Claire to take a random sample of two discs, without replacement from each bag. Write down the mean sample mean. Show that the variance of this mean is 0.583.

Solution 9.16

(a) The mean of the eight numbers, $\mu = 6$
The variance of the eight numbers, $\sigma^2 = 12.5$

(b) Sampling is without replacement, so the mean and variance are given by the formulae on page 135 section (c).

$$\text{mean} = \mu = 6$$

$$\text{variance} = \frac{\sigma^2}{n} \times \frac{N-n}{N-1} = \frac{12.5}{4} \times \frac{4}{7} = 1.79$$

(c) $\mu_{\text{Blue}} = 3$; $\sigma^2_{\text{Blue}} = 0.5$
$\mu_{\text{Red}} = 9$; $\sigma^2_{\text{Red}} = 6.5$

(d) Let B and R be the random samples of size two taken from the blue discs and red discs respectively.

$$\text{The mean sample mean} = E\left[\frac{B+R}{2}\right] = \frac{1}{2}E[B] + \frac{1}{2}E[R]$$

$$= \frac{1}{2} \times 3 + \frac{1}{2} \times 9 = 6$$

The variance of the mean $= \text{Var}\left[\frac{B+R}{2}\right] = \frac{1}{4}[\text{Var}(B) + \text{Var}(R)]$

Now $\text{Var}(B) = \frac{0.5}{2} \times \frac{2}{3} = 0.1667$

and $\text{Var}(R) = \frac{6.5}{2} \times \frac{2}{3} = 2.1667$

So the variance of the mean $= \frac{1}{4}[0.1667 + 2.1667]$

$= 0.583$ (AEB 1988)

***Worked Example 9.17**

Let X and Y be independent random variables each with mean μ but with variances σ_X^2, σ_Y^2 respectively. Show that the weighted mean $wX + (1-w)Y$ where w is a constant, is an unbiased estimator of μ.

Find the value of w which minimises the variance of the weighted mean; find the minimum variance.

Two laboratories made determinations of the amount of an element present in separate collections of 1 g samples of a substance. The following results were presented:

Laboratory	*Mean amount (mg per g substance)*	*Standard error of mean*
A	6.19	0.12
B	6.24	0.24

Use the method developed in the first part of the question to combine the results and give an overall estimate of the mean amount of the element present, together with a standard error.

(OLE)

**Solution 9.17*

$$\begin{aligned} E[wX + (1-w)Y] &= E(wX + Y - wY) \\ &= wE(X) + E(Y) - wE(Y) \\ &= w\mu + \mu - w\mu \\ &= \mu \end{aligned}$$

So the weighted mean is an unbiased estimator of the population mean. As X and Y are independent

$$\text{Var}[wX + (1-w)Y] = w^2\sigma_X^2 + (1-w)^2\sigma_Y^2$$

To find the minimum variance we differentiate the expression for the variance with respect to w and put the result equal to zero.

$$\frac{d}{dw}(w^2\sigma_X^2 + (1-w)^2\sigma_Y^2) = 0$$

$$\Rightarrow \quad 2w\sigma_X^2 - 2(1-w)\sigma_Y^2 = 0$$

$$\Rightarrow w = \frac{\sigma_Y^2}{\sigma_Y^2 + \sigma_X^2}$$

So the minimum variance $= \dfrac{\sigma_Y^4 \sigma_X^2}{(\sigma_Y^2 + \sigma_X^2)^2} + \dfrac{\sigma_X^4 \sigma_Y^2}{(\sigma_Y^2 + \sigma_X^2)^2}$

$$= \frac{\sigma_X^2 \sigma_Y^2 (\sigma_Y^2 + \sigma_X^2)}{(\sigma_Y^2 + \sigma_X^2)^2}$$

$$= \frac{\sigma_X^2 \sigma_Y^2}{\sigma_X^2 + \sigma_Y^2}$$

Final part:

The weighted mean

$$= \frac{(0.24)^2}{(0.24)^2 + (0.12)^2} \,.\, 6.19 + \left(1 - \frac{(0.24)^2}{(0.24)^2 + (0.12)^2}\right) 6.24$$

$= 4.95 + 1.25 = 6.20$

The standard error $= \dfrac{\sigma_X \sigma_Y}{\sqrt{\sigma_X^2 + \sigma_Y^2}}$

$$= \frac{(0.12)\,.\,(0.24)}{\sqrt{(0.12)^2 + (0.24)^2}}$$

$$= 0.107$$

9.9 Exercises

Exercise 9.1

A sample of size 64 is taken from a population of known variance 3.12. If the mean of the sample is 12.96 calculate (i) 95% and (ii) 99% confidence limits for the population mean.

Exercise 9.2

A marketing survey was designed to test whether or not people can tell the difference between butter and a well known brand of margarine. Of the 200 who were given the test, 152 could not tell the difference. Calculate 99% confidence limits for the proportion of people who could tell the difference.

Exercise 9.3

A population is known to follow a Poisson distribution. A sample of size n gives a sample mean of 20. Estimate the size of the sample if the difference in the 95% confidence limits should be less than 2.

Exercise 9.4

It is known that the standard deviation of the time required to solve a certain type of problem is 5 minutes. What is the probability that the mean time of 100 people chosen at random to tackle the problem will differ by more than 1 minute from the mean time taken by all persons tackling the problem?

Exercise 9.5

Two independent surveys of their members were carried out by the motoring organisations, CAR and VAN. Each survey involved asking a randomly chosen sample of their members whether or not they were in favour of raising the speed limit on motorways.

Of the 1000 interviewed by CAR, 550 were in favour of raising the speed limit whilst of the 1500 interviewed by VAN, 720 were in favour. Obtain a 95% confidence interval for the difference in proportions between the members of CAR and the members of VAN who are in favour of raising the speed limit.

Exercise 9.6

Y is a Normal variable with mean 10 and standard deviation 2. A sample of n measurements of the value of Y is taken, and the random variable $\overline{Y}$ is the mean of the n sample values. Write down an expression for the standard deviation of $\overline{Y}$, and find the least value of n for which $P(\overline{Y} > 10.1) < 0.01$. (UCLES)

Exercise 9.7

In the table below, the masses of 40 men are recorded to the nearest kg.

74	87	80	71	77	67	80	83	78	84
75	79	73	79	81	77	89	68	74	93
86	65	82	87	78	92	76	78	73	81
75	73	85	77	73	76	80	83	78	69

Calculate the mean and variance of these masses, showing sufficient working to make clear how your answers are obtained.

Assuming that these masses are a random sample from a population of masses distributed normally with a variance of 40, find 95% confidence limits for the population mean. Explain carefully the meaning to be attached to these limits.

Assuming further that the population mean is 78.4, find the probability that a random sample of eight men from the population will have a total mass exceeding 640 kg. (UCLES)

Exercise 9.8

A random sample of 100 fish was collected from a pond containing a large population of fish; each was marked and returned to the pond. A day later a random sample of 400 fish was taken and 42 were found to be marked. Obtain approximate 90% confidence limits for the proportion of marked fish in the pond. In what way are they approximate? Give similar limits for the size of the population of fish. (SUJB)

Exercise 9.9

The proportion of letters sent by first-class post which are delivered on the next working day after they are posted is p. In order to obtain an estimate of p, 1000 letters were posted at randomly chosen times and places, and their times of arrival were recorded. It was found that 900 were delivered on the next working day after posting. Calculate a 95% confidence interval for p.

Subsequently it is proposed to conduct a larger trial to obtain a more precise estimate of p. Given that p is still near to 0.9, estimate the number of letters that need to be posted in order that the value of p can be determined to within ± 0.005 with 99% confidence. (UCLES)

Exercise 9.10

In a large-scale experiment for comparing two diets A and B, 100 women followed diet A and 80 women followed diet B. From the weight losses of the 100 women who followed diet A it was concluded that the unbiased estimate of the mean weight loss was 2.97 kg and the unbiased estimate of the variance of the weight losses was 1.62 kg^2. From the weight losses of the 80 women who followed diet B the corresponding unbiased estimates were 2.31 kg and 1.55 kg^2, respectively. Stating clearly any assumptions that you make, calculate an approximate 95% confidence interval for the difference between the mean losses in weight for the two diets. (WJEC)

Exercise 9.11

Two firms, A and B, manufacture similar components with a mean breaking strength of 6 kN and 5.5 kN and standard deviations of 0.4 kN and 0.2 kN respectively.

(i) If both distributions are normal, find the probability that one component from manufacturer A and one component from manufacturer B (both components being selected at random) have breaking strengths which differ by at least 0.3 kN.

(ii) If random samples of 100 components from manufacturer A and of 50 from B are tested, find the probability that the mean breaking strength of the components from manufacturer A will be between 0.45 kN and 0.55 kN more than the mean of those from manufacturer B. (UCLES)

Exercise 9.12

A certain brand of beans is sold in tins, the tins being filled and sealed by one of two machines $M1$ or $M2$. From $M1$, the mass of beans in each tin is normally distributed with mean 425 g and standard deviation 25 g and the mass of the tin is normally distributed with mean 90 g and standard deviation 10 g.

(a) Find the probability that the **total** mass of the sealed tin and its beans
  (i) exceeds 550 g.
  (ii) lies between 466 g and 575 g.

(b) Calculate an interval within which approximately 90% of the masses of the filled tins from $M1$ will lie.

  The tins from $M1$ are packed in boxes of 24, the mass of the box being normally distributed with mean 500 g and standard deviation 30 g.

(c) Find the probability that a full box weighs less than 12.75 kg.

A random sample of 10 tins was taken from the production of $M2$ and their total masses (beans and tin), measured to the nearest gram, were as follows

512, 515, 499, 528, 519, 510, 507, 522, 530, 514.

(d) Find a 95% confidence interval for the mean mass of tins of beans, produced on $M2$, assuming that the masses of tins of beans from $M2$ are normally distributed with the same standard deviation as $M1$.

After a delivery of 50 boxes to a supermarket, 150 tins were found to be damaged.

(e) Calculate an approximate 99% confidence interval for the proportion of damaged tins a supermarket might expect to receive. (AEB)

Exercise 9.13

A gardener grew 28 tomato plants. The yield of this population of plants had a standard deviation of 3.2 kg.

(a) Describe how you would use random sampling numbers to take a simple random sample (i.e. the same plant may not be chosen more than once) of size 8 from the population.
(b) What is the standard deviation of the mean yield of the chosen eight plants?
(c) How large a sample would it be necessary to take to make the standard deviation of the mean yield just less than 0.72 kg?

The 28 plants are made up of 21 of strain A and 7 of strain B. The yields of these strains have standard deviation 1.6 kg and 2.5 kg respectively.

(d) $\bar{x}_A$ and $\bar{x}_B$ are the means of simple random samples of size 6 from A and of size 2 from B, respectively. Write down, in terms of $\bar{x}_A$ and $\bar{x}_B$, an expression for $\bar{x}$, the overall mean of a simple random sample of size 6 from A combined with one of size 2 from B.
Find the standard deviation of $\bar{x}$.
(e) Suggest why, in part (d), $\frac{3}{4}$ of the combined sample were taken from A.

(AEB)

***Exercise 9.14**

The random variables X and Y are related to two biased coins A and B in the following way. If when A is tossed a head appears, which it does with probability p, then $X = 1$, and if a tail appears then $X = 0$. The random variable Y is defined in the same way for coin B, for which the probability of a head appearing is $2p$ $(0 < p < 1/2)$.

(i) Show that for all values of λ, the random variable T, defined by

$$T = \lambda X + \frac{1}{2}(1 - \lambda)Y$$

is an unbiased estimator of p.

(ii) Show that $\text{Var}(T) = \lambda^2 p(1-p) + \frac{1}{2}(1-\lambda)^2 p(1-2p)$
(iii) Show that, when $p = 1/3$, the minimum value of $\text{Var}(T)$ is obtained by taking $\lambda = 1/5$.
(iv) Hence obtain, for $p = 1/3$, the least possible value of $P(|\bar{T} - 1/3| > 0.05)$, where $\bar{T}$ is the random variable denoting the mean value of T obtained from 100 tosses of the two coins. (UCLES)

10 Statistical Decision

10.1 Significance Tests

In the last chapter, estimated values of population parameters in the form of confidence intervals were made on the basis of sample information. Due to random factors the point estimate of a population parameter will usually differ from the expected value and it is this difference which will now come under scrutiny. If the difference is *significant*, that is too large, then we may have reason to believe that the true value of the population parameter is different than the expected value. *Significance tests*, based on probability theory, help us to decide whether or not an assumption about a population is true or not.

10.2 Hypotheses

The first step is to make some assumption about the population or populations under consideration. This involves stating a statistical hypothesis about the population(s). The basis of this hypothesis is that there is no significant difference between observed and expected results. It is known as the Null Hypothesis or H_0.

Examples

(i) $H_0 : \mu = 15$ (the population mean is 15)
(ii) $H_0 : \pi_1 - \pi_2 = 0$ (two population proportions are equal)
(iii) H_0 : The observations come from a Binomial distribution.

It is then necessary to set up an Alternative Hypothesis or H_1, which is accepted if the significance test rejects H_0.

Examples

(i) $H_1 : \mu \neq 15$ (the population mean is not 15)
(ii) $H_1 : \pi_1 - \pi_2 < 0$ (one population proportion is larger than the other)
(iii) H_1 : The observations do not come from a Binomial distribution.

10.3 Significance Level and Critical Region

Suppose we wish to test whether an observed sample mean deviates significantly from an assumed population mean μ. For large samples or if the population is known to be normally distributed, the sampling distributions of the means are also normally distributed. If H_0 is true, 95% of sample means will lie within the interval ± 1.96 standard errors of μ. There is a 5% chance therefore that a sample mean lies outside this interval and if our statistical decision is to reject H_0 if this is the case, then there is a probability of 0.05 of being incorrect. That is rejecting H_0 when it is in fact true.

The probability with which we would be willing to risk this type of error is called the *level of significance* of the test. The usual levels, which are specified before testing, are 5% (significant) and 1% (highly significant) but other levels are sometimes appropriate.

In order to reject H_0 at the 5% level of significance, the sample mean should lie in one of the two tails of the Normal distribution curve shown in Fig. 10.1. This region is called the *critical region* and the boundary values are called *critical values*. A significance test of this nature is called a *two-tailed test.*

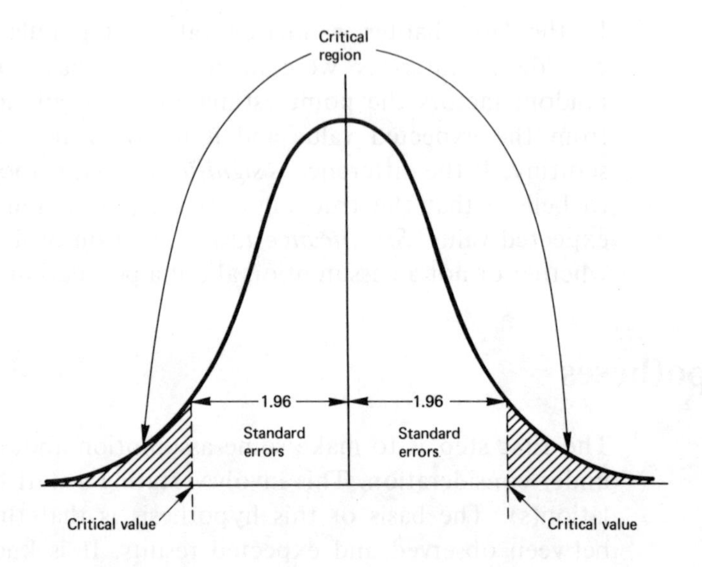

Figure 10.1

10.4 Test Statistic

To calculate the difference between the observed value and the assumed population parameter in standard error units we use the *test statistic*

$$z = \frac{\text{observed value} - \text{assumed population parameter}}{\text{standard error}}$$

For a population with known variance σ^2, the test statistic for the difference between the sample mean $\overline{x}$, and the population mean μ, is

$$z = \frac{\overline{x} - \mu}{\sigma/\sqrt{n}}$$

Once z is calculated a decision can then be made. We can reject H_0 in favour of H_1 if z is in the critical region, or otherwise we accept H_0.

10.5 Summary of Procedure for Significance Testing

(i) State Null and Alternate Hypotheses.
(ii) Choose a level of significance.
(iii) Select a probability distribution.
(iv) Choose a critical region.
(v) Carry out computations.
(vi) Make the statistical decision.

Worked Example 10.1

A Normal distribution with known standard deviation of 3 units is thought to have a mean of 20 units. A random sample of 9 items from this distribution gives a sample mean of 18.2. Does this support the view that the population mean is 20 units?

Solution 10.1

$H_0 : \mu = 20$
$H_1 : \mu \neq 20$

In the majority of cases it is enough to show if a test statistic is significant and therefore a 5% level of significance is adequate.

Under H_0, the distribution of $\overline{X}$ the mean of samples of size 9 is normally distributed with mean = 20 and standard error = 3/3 = 1.

The critical values of the test statistic z, are $\pm$ 1.96. Now

$$z = \frac{\overline{x} - \mu}{\sigma/\sqrt{n}}$$

$$= \frac{18.2 - 20}{1}$$

$$= -1.8$$

Clearly z does not lie in the critical region so the evidence supports the view that the population mean is 20 units.

10.6 One-tailed and Two-tailed Tests

In the worked example we would have rejected H_0 if the sample mean fell in either of the two rejection tails. There are occasions however when H_1 suggests a bias in one particular direction only. For a 5% level of significance in this instance the rejection region is a 5% tail on one particular side only.

10.7 Critical Values

The test statistic z is from a standardised Normal distribution and the critical values are as follows:

	Two-tailed test	*One-tailed test*
5%	± 1.96	1.645 or − 1.645
1%	± 2.575	2.33 or − 2.33
0.1%	± 3.3	3.1 or − 3.1

Worked Example 10.2

In Worked Example 10.1, test H_0 against the Alternative Hypothesis that the population mean is less than 20.

Solution 10.2

$H_0 : \mu = 20$
$H_1 : \mu < 20$

Proceeding as before, the standard error = 1 and for a 5% level of significance the critical value is now -1.645. The test statistic z is again -1.8.

This time z does lie in the critical region and so we reject H_0 in favour of H_1.

Worked Example 10.3

A die is thrown 300 times and 64 sixes were observed. Test at the 1% level of significance whether this result provides evidence that the die is biased towards a six.

Solution 10.3

$H_0 : \pi = 1/6$
$H_1 : \pi > 1/6$ (one tail)

Under H_0, the distribution P of the proportion of sixes is normally distributed

$$\text{with mean} = \pi = \frac{1}{6} = 0.167$$

$$\text{and standard error} = \sqrt{\frac{\pi(1-\pi)}{n}} = \sqrt{\frac{\frac{1}{6}\cdot\frac{5}{6}}{300}} = 0.0215$$

For a 1% level of significance (one-tail) the critical value is 2.33. The test statistic

$$z = \frac{p - \pi}{\sqrt{\frac{\pi(1-\pi)}{n}}} = \frac{0.213 - 0.167}{0.0215}$$

$$= 2.14$$

As z is less than the critical value we have no evidence that the die is biased towards a six.

10.8 Type I and Type II Errors

The decision procedure gives rise to two possible errors.

Type I The rejection of H_0 when in fact it is true. No observation is impossible and the probability of this error is the same as the level of significance.

Type II The acceptance of H_0 when it is false. Unlike a Type I error which is a constant only dependent on the level of significance, this error is dependent on which alternative hypothesis is actually true.

Figure 10.2 shows a representation of the two types of errors. C_1 and C_2 are critical values for H_0.

Minimising the errors of decision is not an easy task and from the diagram it should be seen that any attempt to decrease one error increases the other. The only way to decrease both types of errors is to increase the sample size.

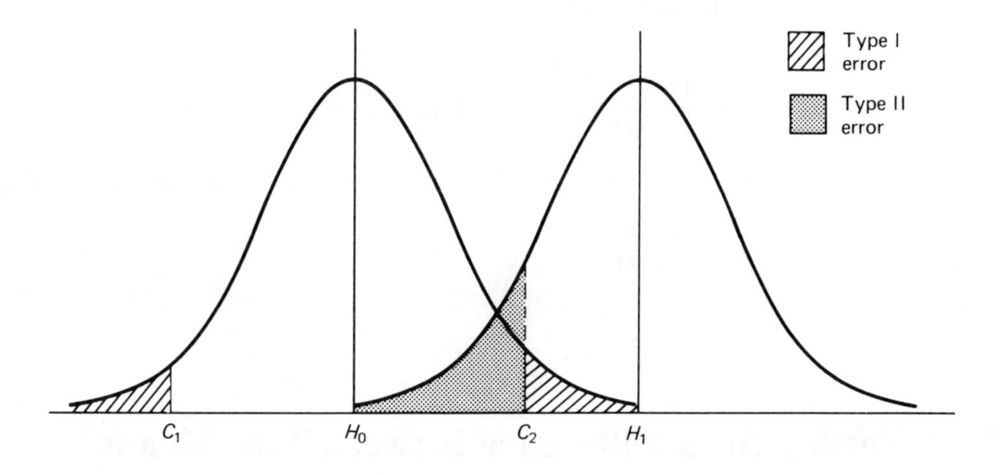

Figure 10.2

Worked Example 10.4

It is suspected that a die is biased towards a six with a probability of 0.3. To test whether the die is biased in this manner, it is rolled 120 times. The die is accepted as being biased if more than 26 sixes are thrown. Calculate the Type I error and also the Type II error if it is accepted that the die is biased when in fact it is fair.

Solution 10.4

$H_0 : \pi = 0.3$
$H_1 : \pi = 0.167$

Under H_0, the sampling distribution is approximately normally distributed with mean 0.3 and standard error $\sqrt{\dfrac{0.3\,.\,0.7}{120}}$.

Allowing for a continuity correction factor, we standardise 26.5/120

$$z = \frac{\dfrac{26.5}{120} - 0.3}{\sqrt{\dfrac{0.3\,.\,0.7}{120}}} = -1.89$$

The probability of a Type I error is the probability of 26 or less sixes thrown

$= P(Z < -1.89)$
$= 0.0294$

For the Type II error H_1 is true therefore the sampling distribution is approximately normally distributed with mean 0.167 and standard error of

$$\sqrt{\frac{\dfrac{1}{6}\cdot\dfrac{5}{6}}{120}} = 0.034$$

Standardising 26.5/120

$$z = \frac{\frac{26.5}{120} - 0.167}{0.034} = 1.58$$

The probability of a Type II error is the probability of more than 26 sixes thrown

$= P(Z > 1.58)$
$= 0.0571$

10.9 Testing for a Difference between Two Means

In the previous chapter we stated that the sampling distribution of the difference between the sample mean $\overline{X} - \overline{Y}$ is distributed with mean $\mu_X - \mu_Y$ and variance $\frac{\sigma_X^2}{n_X} + \frac{\sigma_Y^2}{n_Y}$ and that this distribution will be Normal provided that the parent populations are Normal or the sample sizes are large. When testing for a difference between two means, H_0 is the assumption that there is no difference between the population means. Therefore the difference between the two sample means under this hypothesis is normally distributed with a mean of 0 and a standard error of $\sqrt{\frac{\sigma_X^2}{n_X} + \frac{\sigma_Y^2}{n_Y}}$. The test statistic is now $z = \frac{(\overline{x} - \overline{y}) - 0}{\sqrt{\frac{\sigma_X^2}{n_X} + \frac{\sigma_Y^2}{n_Y}}}$

where $\overline{x}$ and $\overline{y}$ are the respective sample means.

Worked Example 10.5

An investigation into the ages at which a person acquired a full driving licence in the country of Carmania produced the following information (ages are decimals, not years and months):

> In the state of Mobilia, the average age of 160 people was 21.76 years with a standard deviation of 2.9 years.
> In the state of Pedestria, the average age of 120 people was 22.82 years with a standard deviation of 2.4 years.

Do these figures suggest that there is a significant difference between the average ages that people from the states of Mobilia and Pedestria acquire a driving licence?

Solution 10.5

Let M represent the age of a person acquiring a licence in Mobilia and let P represent the age of a person acquiring a licence in Pedestria.

$H_0 : \mu_M - \mu_P = 0$ there is no difference between the mean ages.
$H_1 : \mu_M - \mu_P \neq 0$ there is a difference between the mean ages.

Then $\overline{M} - \overline{P}$ under H_0 is normally distributed with a mean of 0 and a standard error of $\sqrt{\frac{\hat{s}_M^2}{n_M} + \frac{\hat{s}_P^2}{n_P}}$.

We are given s_M and s_P the sample standard deviations and although in practice these are used directly as estimates for σ_M and σ_P, we shall for completeness calculate unbiased estimates for the population variances.

$$\hat{s}_M^2 = \frac{160}{159} \cdot 2.9^2$$

and $$\hat{s}_P^2 = \frac{120}{119} \cdot 2.4^2$$

Using these as the best estimates of the population variances the standard error

$$= \sqrt{\frac{2.9^2}{159} + \frac{2.4^2}{119}}$$

$$= 0.318$$

For a 5% level of significance and a two-tailed test, the critical values are ± 1.96.

$$\text{The test statistic } z = \frac{(\bar{x}_M - \bar{x}_P) - 0}{0.318}$$

$$= \frac{(21.76 - 22.82) - 0}{0.318}$$

$$= -3.33$$

As z is less than -1.96 the result is significant and we reject H_0 in favour of H_1. That is we assume that there is a difference between the two population means.

Note: Using this method we can actually state the level of significance. For a two-tailed test $P(|Z| > 3.33)$ is approximately 0.001 so we can say that the result is very highly significant at a 0.1% level.

10.10 Tests for Differences between Proportions

In a similar manner we can test whether there is any significant difference between two population proportions using the fact that $P_1 - P_2$ for large samples is approximately normally distributed with mean $(p_1 - p_2)$ and standard error $\sqrt{\frac{p_1(1-p_1)}{n_1} + \frac{p_2(1-p_2)}{n_2}}$.

When testing for a difference between two proportions, H_0 is the assumption that there is no significant difference between the two proportions therefore the difference between the two sample proportions is distributed with a mean of 0 and a standard error of $\sqrt{\frac{p(1-p)}{n_1} + \frac{p(1-p)}{n_2}}$ where $p = \frac{n_1 p_1 + n_2 p_2}{n_1 + n_2}$, the pooled estimate of the population proportion.

Worked Example 10.6

A microcomputer dealer Mr E Prom, sells two popular makes of home computer, the Romulus and the Ramus. In the first year of sales, 9 of the 200 Romulus computers sold were returned for repair whilst 16 of the 250 Ramus computers sold were returned for repair. Is there any evidence that the Romulus computer was more reliable?

Solution 10.6

$H_0 : \pi_1 - \pi_2 = 0$	the proportion of unreliable machines of the two types are the same.
$H_1 : \pi_1 - \pi_2 < 0$	the proportion of unreliable Romulus computers is less than the proportion of unreliable Ramus computers.

Under H_0 the difference in the sample proportions is distributed with mean 0 and standard error $\sqrt{\frac{\pi(1-\pi)}{n_1} + \frac{\pi(1-\pi)}{n_2}}$.

$$\text{The best estimate for } \pi \text{ is } p \text{ where } p = \frac{n_1 p_1 + n_2 p_2}{n_1 + n_2}$$

$$= \frac{9 + 16}{200 + 250}$$

$$= 0.056$$

$$\text{So the standard error} = \sqrt{\frac{0.056\,.\,0.944}{200} + \frac{0.056\,.\,0.944}{250}}$$

$$= 0.022$$

This is a one-tailed test so for a 5% level of significance, the critical value is -1.645.

$$\text{The test statistic } z = \frac{(p_1 - p_2) - 0}{0.022}$$

$$= \frac{(0.045 - 0.064) - 0}{0.022}$$

$$= -0.86$$

As z is not less than -1.645, then the observed difference is not in the critical region and we have no reason to reject H_0. Therefore there is no evidence that the Romulus computer is more reliable.

10.11 Paired Comparison Tests for Large Samples

Two related sets of sample data can be tested to see if there is any significant difference between them by considering the value of the difference for each related pair. If we let D be the distribution of the differences then H_0 would be that there is no significant difference between the pairs and the appropriate test would be that the mean of D is distributed with a mean of 0. As the pairs are not independent, we cannot calculate a standard error from the variances of the two distributions. Instead we have to use the variance of the differences. For questions involving large samples, this is usually given.

Worked Example 10.7

An examination consists of two equally weighted papers. The difference between the mean marks of 100 candidates is 2.7 and the standard deviation of the differences is 12.4. Would we be justified in assuming that the mean marks for the two papers are the same?

Solution 10.7

$H_0 : \mu = 0$
$H_1 : \mu \neq 0$

The unbiased estimator of the population variance is $100\,.\,12.4^2/99$ so the standard error of the differences is $\frac{12.4}{\sqrt{99}} = 1.25$.

For a 5% level of significance, two-tailed test, the critical values are $\pm$ 1.96.

$$\text{The test statistic } z = \frac{2.7 - 0}{1.25} = 2.16$$

As $z > 1.96$ the value lies in the critical region and therefore we accept H_1 at a 5% level of significance. The exact level of significance by using the Normal distribution tables is 3.08%.

Worked Example 10.8

A random sample of 100 measurements of a variable X are taken and the following results are obtained:

$$\sum_{1}^{100} (x - 20) = 125 \qquad \sum_{1}^{100} (x - 20)^2 = 780$$

Test the hypothesis at the 5% level of significance that the mean value of the distribution is 22.

Solution 10.8

$H_0 : \mu = 22$
$H_1 : \mu \neq 22$

From the information above we need to calculate a sample mean and an unbiased estimator of the population variance.

$$\text{Sample mean} \quad \bar{x} = 20 + \sum_{1}^{100} \frac{(x - 20)}{100} = 21.25$$

$$\text{Sample variance } s^2 = \sum_{1}^{100} \frac{(x - 20)^2}{100} - \left(\sum_{1}^{100} \frac{(x - 20)}{100}\right)^2 = 7.8 - 1.563 = 6.237$$

The unbiased estimate of the population variance $\hat{s}^2$,

$$= \frac{100 \,.\, 6.237}{99} = 6.3$$

Under H_0, the sampling distribution of the means is normally distributed with mean of 22 and standard error of $\sqrt{6.3/100}$. For a 5% level of significance, two-tailed test, the critical values are $\pm$ 1.96.

$$\text{The test statistic } z = \frac{21.25 - 22}{\sqrt{6.3/100}} = -2.99$$

This result is clearly highly significant (0.28%) and we reject H_0 in favour of H_1.

Worked Example 10.9

A survey of 144 weddings in Shoreton showed that 54 took place in a Registry Office. Does this data agree with the ratio of Church to Registry Office weddings of 2 : 1?

Solution 10.9

$H_0 : \pi = 2/3$
$H_1 : \pi \neq 2/3$

Under H_0, the distribution P of the sample proportions is normally distributed with mean 2/3 and standard error $\sqrt{\dfrac{\frac{2}{3}\cdot\frac{1}{3}}{144}} = \dfrac{\sqrt{2}}{36}$

Testing at the 5% level of significance, the critical values for a two tailed test are ± 1.96.

$$\text{The test statistic } z = \frac{\dfrac{(144-54)}{144} - \dfrac{2}{3}}{\sqrt{2}/36}$$

$$= \frac{-1}{24} \Big/ \frac{\sqrt{2}}{36} = \frac{-3}{2\sqrt{2}}$$

$$= -1.06$$

Clearly we have no reason to reject H_0 and we accept that there is no evidence to suggest that the ratio of Church to Registry Office weddings is not 2 : 1.

Worked Example 10.10

In Taxville, a random survey of n cars shows that 27% have no valid road tax. What is the value of n if this proportion is to be significantly different from a population proportion of 30% at a level of 5%.

Solution 10.10

Under a Null Hypothesis that the population proportion is 0.3, the distribution of sample proportions P is normally distributed with a mean of 0.3 and a standard error of $\sqrt{0.3 \,.\, 0.7/n}$.

For 0.27 to be significant at the 5% level, the sample proportion must be at least 1.96 standard errors from the assumed population proportion.

$$(0.3 - 0.27) > 1.96 \sqrt{\frac{0.3 \,.\, 0.7}{n}}$$

$$n > 1.96^2 . \frac{0.3 \,.\, 0.7}{0.03^2}$$

$$n > 896.4$$

So the value of n is 897.

Worked Example 10.11

Two roads lead into Tartanland, the High Road and the Low Road. It is suggested that the proportion of motorists who take either road is the same and in order to test this hypothesis, two simple procedures are set up. Both procedures reject the hypothesis that the proportions are the same if:

Procedure 1. A random sample of four motorists are all taking the same road.
Procedure 2. Of a random sample of seven motorists, at least six are taking the same road.

Show that each of the procedures has the same Type I error, that is accepting that the proportions are different when in fact they are the same. Which of the procedures leads to a smaller Type II error, that is accepting that the proportions are the same when in fact the probability that a motorist takes the High Road is 3/4?

Solution 10.11

Under the Null Hypothesis, the probability of four motorists all taking the same road is $2 \cdot \left(\frac{1}{2}\right)^4 = \frac{1}{8}$.

The probability that of seven motorists chosen at random at least six are taking the same road is $2 \cdot \left[7 \cdot \left(\frac{1}{2}\right)^6 \cdot \left(\frac{1}{2}\right) + \left(\frac{1}{2}\right)^7\right] = \frac{1}{8}$.

Both these are the probabilities of rejecting H_0 when in fact it is true. Hence the probability of a Type I error under both procedures is the same.

If H_1 is true ($\pi = 3/4$ for the High Road) then the probability of a Type II error under procedure 1 is $1 - \left[\left(\frac{1}{4}\right)^4 + \left(\frac{3}{4}\right)^4\right]$

$$= 0.68$$

A Type II error under procedure 2 is

$$1 - \left[\left(\frac{1}{4}\right)^7 + 7\left(\frac{1}{4}\right)^6\left(\frac{3}{4}\right) + 7\left(\frac{1}{4}\right)\left(\frac{3}{4}\right)^6 + \left(\frac{3}{4}\right)^7\right]$$

$= 0.55$

So procedure 2 leads to the smaller Type II error.

Worked Example 10.12

The number of particles emitted by a radioactive source in a time interval of 1 second has a Poisson distribution with mean 1.5. On the assumption that successive time-intervals are independent, what distribution is followed by the number of particles emitted in a minute, and what distribution may be used to approximate this distribution?

What is the probability that the number of particles emitted in one minute is less than 70?

The number of particles recorded from a second source in a minute is 75. Use an approximate distribution to obtain a 95% confidence interval for the mean number emitted in a minute. Use your confidence interval to test whether the second source is emitting particles at the same rate as the first. (OLE)

Solution 10.12

As the time intervals are independent, we can use the additive property of the Poisson distribution. Therefore the number of particles emitted/minute follows a

Poisson distribution with mean 90, and variance of 90. We can use the Normal distribution as an approximation to the discrete Poisson distribution. If X is the distribution of the number of particles emitted/minute we require $P(X < 70)$.

Allowing for a continuity correction factor, we standardise 69.5.

$$\text{The standard score } z = \frac{69.5 - 90}{9.49}$$

$$= -2.16$$

$$\text{So} \qquad P(X < 70) = P(Z < -2.16)$$

$$= 0.0154$$

Using a Normal approximation, the 95% confidence limits are

$$= 75 \pm 1.96 \,.\, 8.66$$

$$= 75 \pm 16.97$$

So the confidence interval is (58.03, 91.97).

If the Null Hypothesis is that the mean number of emissions/minute is 90, then this figure lies in the 95% confidence interval and so on this basis the difference between observed and expected results is not significant at the 5% level.

Worked Example 10.13

The mean number of letters delivered to the School Office and to the Principal's Office were recorded over varying periods of time:

	No. of days recorded	Mean no. of letters	Standard deviation
School Office	63	48.1	6.24
Principal's Office	52	44.6	8.56

Is there a significant difference between the means? (Use a 5% level of significance.) (SUJB)

Solution 10.13

Let μ_s be the population mean number of letters delivered to the School Office daily and μ_p be the corresponding mean for the Principal's Office.

$H_0 : \mu_s - \mu_p = 0$
$H_1 : \mu_s - \mu_p \neq 0$

If $\overline{S}$ and $\overline{P}$ are the sampling distributions of the means, then under H_0, the distribution of $\overline{S} - \overline{P}$ is normally distributed with a mean of 0 and a standard error of $\frac{\hat{s}_s^2}{n_s} + \frac{\hat{s}_p^2}{n_p}$

where $\hat{s}_s^2$ and $\hat{s}_p^2$ are unbiased estimators of the population variances.

$$\hat{s}_s^2 = \frac{63\,.\,(6.24)^2}{62}$$

$$\text{and} \qquad \hat{s}_p^2 = \frac{52\,.\,(8.56)^2}{51}$$

So the standard error $= \sqrt{\frac{(6.24)^2}{62} + \frac{(8.56)^2}{51}}$

$= 1.44$

The test statistic $z = \frac{(\bar{x}_s - \bar{x}_p) - 0}{1.44}$

$= \frac{48.1 - 44.6}{1.44}$

$= 2.43$

Since z is much greater than 1.96 the result is highly significant (1.5%) and we would reject H_0 in favour of H_1.

Worked Example 10.14

An urn contains a large number of marbles which are either black or gold in colour. To test the hypothesis that there is an equal proportion of black and gold marbles in the urn, a sample of size 25 is selected at random with replacement. The hypothesis is accepted if the number of gold marbles in the sample lies between 10 and 15 inclusive, otherwise the hypothesis is rejected. Estimate the probability of rejecting the hypothesis when it is actually correct. If you wanted to test the hypothesis that there was a greater number of gold marbles than black in the urn, at the 5% level of significance, using a sample of 64 selected at random with replacement, what decision rule would you adopt? (L)

Solution 10.14

The first part of the question is asking us to find the Type I error. If we let X be the number of gold marbles in a sample then H_0 is accepted if $10 \leqslant X \leqslant 15$. Under H_0, X follows a binomial distribution with mean 12.5 and variance 6.25. Using the Normal approximation to the binomial and standardising 9.5 and 15.5

$$z_1 = \frac{9.5 - 12.5}{\sqrt{6.25}}$$

$$= -1.2$$

$$z_2 = \frac{15.5 - 12.5}{\sqrt{6.25}}$$

$$= 1.2$$

So the probability of a Type I error

$$= 1 - P(-1.2 < Z < 1.2)$$

$$= 1 - 2 \times 0.385$$

$$= 0.23$$

The decision rule required necessitates finding a numerical value which if the number of gold balls drawn from the urn exceeds, will provide evidence that the number of gold balls exceeds the number of black balls. We require therefore to reverse the process of the first part.

For a one-sided distribution where the probability of a Type I error is 5%, the corresponding standard score is 1.645.

So $1.645 = \frac{X - 32}{\sqrt{16}}$ $\left[\text{Note: } npq = 64 \times \frac{1}{2} \times \frac{1}{2}\right]$

$$X = 32 + 1.645 \times 4$$
$$= 32 + 6.58$$
$$= 38.6$$

Allowing for the continuity correction factor, the decision rule would be to accept the hypothesis that there is a greater number of gold balls than black balls if the number of gold balls drawn exceeds 38.

Worked Example 10.15

Table 10.1 gives the frequency distribution of the masses of 113 animals, recorded to the nearest gram.

Table 10.1

Mass (g)	41	42	43	44	45	46
Frequency	3	17	36	38	15	4

Calculate the mean and variance of the given frequency distribution.

(a) If a normal distribution is to be fitted to the data, what values would you estimate for its mean and variance?
(b) What would be the estimated mean and variance of the distribution of the mean mass of 100 such animals selected at random?
(c) If a random sample of 50 animals is taken from a normal distribution with the same variance as the normal distribution in (a), what value would you estimate for the variance of the difference between the mean of such a sample and the mean of one of 100 from the distribution in (a)?
(d) Use a 95% confidence interval for the mean of the normal distribution in (a) to test whether this mean is significantly above 43.3 g. (OLE)

Solution 10.15

Using coding techniques we will find $\bar{x} = 43.5$

$$s^2 = 1.188$$

(a) Estimation of the mean $= 43.5$

Estimation of the variance
$$= \frac{n \,.\, s^2}{n-1}$$
$$= \frac{113}{112} \,.\, 1.188$$
$$= 1.199$$

(b) Estimation of the mean mass of 100 animals selected at random

$$= 43.5$$

Variance of the sampling distribution of the mean mass of 100 animals

$$= \frac{1.199}{100}$$
$$= 0.012$$

(c) Variance of the difference between the two sample means

$$= \frac{\hat{s}^2}{n_1} + \frac{\hat{s}^2}{n_2}$$

$$= \hat{s}^2 \left(\frac{1}{50} + \frac{1}{100}\right)$$

$$= 1.2 \times 0.03$$

$$= 0.036$$

(d) $H_0 : \mu = 43.3$
$H_1 : \mu > 43.3$

95% confidence limits for the population mean

$$= \bar{x} \pm 1.96 . \frac{\sqrt{1.2}}{\sqrt{113}}$$

$$= 43.5 \pm 0.202$$

As 43.3 lies within this interval we have no reason to reject H_0.

10.12 **Exercises**

Exercise 10.1

A population is known to be normally distributed with a variance of 0.2 unit^2. Nine values chosen at random from this population are as follows:

17.6, 19.0, 18.6, 20.1, 19.2, 18.4, 18.0, 19.3, 20.6.

Test at the 5% level of significance that the population mean is 19.2 units.

Exercise 10.2

The recovery rate from a particular disease is 50%. A new drug tried out on 200 people suffering from the disease resulted in 112 people recovering. Test at the 5% level of significance whether the new drug is effective.

Exercise 10.3

Mr Triagain and Mr Underpass are two driving instructors for the Passalot driving school. Given that of 60 pupils tutored by Mr Triagain, 40 pass their test at the first attempt and of 70 pupils tutored by Mr Underpass, 43 pass at the first attempt. Is there any evidence of a difference in these results?

Exercise 10.4

A flour mill uses two machines to fill bags with 1 kg of flour. Over a period of time the weights of the bags were recorded and the results were as in Table 10.2.

Table 10.2

	Machine A	*Machine B*
Mean weight	998 g	1001 g
Standard deviation	6 g	9.5 g
Number of bags	80	120

Test at the 5% level of significance if there is any difference between the mean weights of bags of flour produced by the two machines.

Exercise 10.5

A bag is known to contain either

(a) 3 red balls and 7 black balls, or
(b) 6 red balls and 4 black balls.

Letting the Null Hypothesis H_0 be option (a) and the Alternative Hypothesis be option (b), the hypotheses are tested against each other by drawing out 2 balls from the bag. If at least one ball is black the Null Hypothesis is accepted, otherwise it is rejected. Find the Type I and Type II errors.

Exercise 10.6

The marks of 200 candidates in an examination were as in Table 10.3. Find the mean and standard deviation of this data. If this data is a random sample from a large population test the hypothesis that the population mean is 50.

Table 10.3

Marks	1-10	11-20	21-30	31-40	41-50
Frequency	4	8	18	32	46
Marks	51-60	61-70	71-80	81-90	91-100
Frequency	48	24	12	6	2

Exercise 10.7

A machine is designed to produce grub screws with a nominal length of 5 mm. A random sample of 400 screws produced by the machine is found to have a mean length of 5.008 mm and a standard deviation of 0.072 mm. Estimate the S.E. of the mean, and obtain an approximate 95% confidence interval for the mean of the whole output of this machine.

State giving your reason, whether, on the evidence of this sample, the mean of the whole output differs significantly at the 5% level from the nominal value of 5 mm. (L)

Exercise 10.8

Experimental data concerning a variable X, which measures the reliability of a certain electronic component, is as follows:

$\Sigma x = 1164.2$, $\Sigma x^2 = 13\,911.6$, $n = 100$. Calculate the sample mean and standard deviation from these figures. Explain whether, on the evidence of this sample, you would reject the hypothesis that the mean value of X is 12.

Figures collected over a long period have established that the mean and standard deviation of X are 12 and 2 respectively. After a change in the manufacturing process it is expected that the mean would have been increased, but it may be assumed that the standard deviation remains equal to 2. A sample of n values of X is taken, with sample mean m; if m is greater than some critical value it will be accepted that the mean has in fact increased, but if m is less than the critical value the increase is not established. State carefully appropriate null and alternative hypotheses for this situation, and find, in terms of n, the critical value for a 1% significance level. (UCLES)

Exercise 10.9

(a) A certain type of battery for calculators is said to last for 2000 hours. A sample of 200 of these batteries was tested; the mean life was 1995 hours and the standard deviation of the lives was 25.5 hours. Use these data to test the hypothesis that the population mean life is 2000 hours against the alternative hypothesis that it is less than 2000 hours. State what level of significance you are using in your test.

(b) Two types of battery were compared for the length of time they lasted. The data obtained are summarised in Table 10.4.

Table 10.4

Battery type	*No. tested*	*Sample mean*	*Sample S.D.*
A	200	1995	25.5
B	150	2005	32.8

Test the hypothesis that the population from which these samples were drawn have equal means against the alternative hypothesis of unequal means. State the level of significance you are using in your test. (UCLES)

Exercise 10.10

In the course of a survey concerning the proportion of left-handed children the figures in Table 10.5 were obtained from two schools. Show that an approximate 95% confidence interval for the population proportion, p, of left-handed children derived from the data from School 1 is $0.25 < p < 0.32$, and calculate a corresponding interval for School 2.

Table 10.5

	Number of children	*Proportion left-handed*
School 1	620	0.284
School 2	475	0.341

Explain briefly what is wrong with the following argument: 'Since these two confidence intervals overlap, we cannot reject, at the 5% significance level, the hypothesis that the populations from which the children in the two schools are samples each have the same proportion of left-handed children'.

Calculate the overall proportion of left-handed children in both schools, and show that the observed difference in proportions is significant at the 5% level.

(UCLES)

Exercise 10.11

Tests by a tyre company on their 'Special' tyres under harsh conditions have shown that 9% of the tyres have a safe life of under 30 000 km while 20% have a safe life of more than 50 000 km. The safe life of the tyres can be modelled by a normal distribution.

A car is fitted with four new 'Special' tyres and is driven under the same harsh conditions. Find, to 2 decimal places, the probability that by the time the car has covered 45 000 km, exactly three of the four original tyres will have had to be replaced.

The engine of the car has an essential component C. In a journey of D km the number of times that C has to be replaced has a Poisson distribution with mean $\frac{D}{18\,000}$. Find, to 2 decimal places, the probability that the car will complete 45 000 km, under harsh conditions, without needing a tyre change and having had no more than 4 replacements of the component C.

A new process is introduced into the manufacture of 'Special' tyres. A random sample of 100 tyres then produced is selected and tested under the same harsh conditions. It is found that their mean safe life is 43 900 km. Test, at the 5% level of significance, whether or not the introduction of the new process has been successful. State clearly your null and alternative hypotheses. (L)

Exercise 10.12

A manufacturer of dice makes fair dice and also slightly biased dice which are to be used for demonstration purposes in the teaching of probability and statistics. The probability distributions of the scores of the two types of die are shown in Table 10.6. Calculate the expectation and variance of the score for each type of die.

Table 10.6

Score	*1*	*2*	*3*	*4*	*5*	*6*
Probability for fair dice	1/6	1/6	1/6	1/6	1/6	1/6
Probability for biased dice	1/10	1/10	1/5	1/5	1/5	1/5

Unfortunately, some dice have been made without distinguishing marks to show whether they are fair or biased. The manufacturer decides to test the dice as follows: each die is thrown 100 times and the mean score $\bar{x}$ is calculated; if $\bar{x} > 3.7$, the die is classified as biased but, if $\bar{x} \leqslant 3.7$, it is classified as fair. Find the probability that the fair die is wrongly classified as biased as a result of this procedure.

(UCLES)

Exercise 10.13

A blood test leads to a measure, X, which is known to be abnormally high among sufferers from Paget's disease. From hospital records it is known that, for sufferers it is normally distributed with mean 16.5 and standard deviation 6.0, while among the rest of the population it is normally distributed with mean 7.4 and standard deviation 3.0.

As a screening test it is decided to diagnose (provisionally) as suffering from the disease those people tested who have an X value greater than 14.

Taking the null hypothesis to be that the person tested does not suffer from Paget's disease, what is the risk of making

(a) a Type 1 error,
(b) a Type 2 error?

What range of values of X should lead to a provisional diagnosis of Paget's disease to make the risk of

(c) a Type 2 error equal to 5%,
(d) a Type 1 error equal to the risk of a Type 2 error.

Paget's disease is believed to affect 5% of people over 60 years of age. It is proposed that all people over 60 should be screened, and those whose X value exceeds 14 are further investigated.

(e) What proportion of people over 60 would be further investigated?
(f) Of those further investigated what proportion would have the disease?

(AEB 1989)

***Exercise 10.14**

The lifetime, T, in hours of a certain type of electric lamp is a random variable with distribution

$$f(t) = Ae^{-t/1200}, 0 \leqslant t < \infty,$$
$$= 0, t < 0$$

Find the value of A and show that the mean and standard deviation of T are both 1200 hours.

To test the reliability of the production a random sample of 40 bulbs was tested and found to have a mean life of 1020 hours. Does this indicate at the 5% level of significance that the batch from which the sample was taken was substandard? (SUJB)

***Exercise 10.15**

A given circle has equation $x^2 + y^2 = r^2$. A point P is selected at random on the diameter from $(-r, 0)$ to $(r, 0)$, uniformly distributed over that diameter, and the chord through P parallel to the y-axis is drawn. Show that the probability that the length of the chord exceeds the radius of the circle is $\sqrt{3}/2$.

Margaret believes that the random number tables she is using to select such points are biased, so that points near the centre of the circle are more likely to be chosen. She chooses 500 points independently, and finds that 450 of the chords are longer than the radius. Test her belief. (OLE)

11 Small Sampling

11.1 Definition of a Small Sample

The techniques of Chapters 9 and 10 were appropriate in general to large samples of size $n > 30$. In these cases, the sampling distributions of many statistics were approximately normal. The approximation is better the larger the value of n. If $n < 30$, the sample is said to be a *small sample*, and the Normal approximation is not good. The sampling distribution of a statistic is only normally distributed if (i) the population distribution is normally distributed and (ii) the population variance σ^2 is known. Since σ^2 usually isn't known it has to be approximated by $\hat{s}^2$ and small sampling techniques must then be used. The appropriate distribution is the Student's t distribution as defined in Section 11.2. This distribution is in fact true for all values of n.

11.2 Student's *t* Distribution

The t test statistic is defined by

$$t = \frac{\bar{x} - \mu}{\hat{s}/\sqrt{n}}$$

The probability density function of t is given by

$$f(x) = T_\nu \left(1 + \frac{x^2}{\nu}\right)^{-\frac{(\nu + 1)}{2}} \qquad -\infty < x < \infty$$

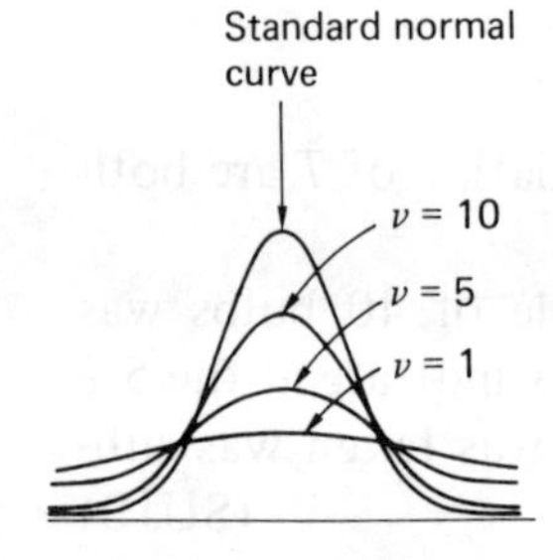

Figure 11.1

T_ν is a constant depending on ν where ν is a parameter which takes integer values and is the number of degrees of freedom of the distribution. In fact, $\nu = n - 1$. See Section 11.3. It can be seen in Fig. 11.1 that as $\nu \to \infty$, the t distribution approaches the Normal distribution.

11.3 Degrees of Freedom

When calculating the value of a statistic, it may be necessary to use data from a sample, or certain population parameters. If the population parameters are not known, they have to be estimated from the sample. If k is the number of population parameters that have to be estimated, then the number of degrees of freedom ν for a sample of size n is given by $\nu = n - k$. In the case of calculating t, $\bar{x}$ and $\hat{s}$ can be calculated from the sample data, whereas μ must be estimated. Hence $k = 1$.

11.4 Use of t-Tables

The t distribution tables are given in Table 16.2. The table given is for one-tailed tests. For example, if $\nu = 8$, and the 95% significance level for a two-tailed test is required, look at the column headed 0.025. The significance value is 2.306. Hence the probability that $-2.306 \leqslant t \leqslant 2.306$ is 95%.

11.5 Significance of a Single Mean, Confidence Intervals

In order to carry out this type of test, the best estimate of the population mean is the mean of the sample, and the best estimate of the population variance is $\dfrac{ns^2}{n-1}$.

The confidence intervals for μ are given by

$$\bar{x} \pm t_\nu \frac{\hat{s}}{\sqrt{n}}$$

where t_ν denotes the critical value for ν degrees of freedom.

Worked Example 11.1

A sample of 12 measurements of the mass of lead weights gave a mean of 25.6 g and a standard deviation of 0.084 g. What are the 95% confidence limits for the mean mass of the lead weights?

Solution 11.1

Best estimate for $\mu = 25.6$

$$\hat{s}^2 = \frac{n}{n-1} s^2 = \frac{12}{11} \times 0.084^2 = 0.007697$$

$\Rightarrow \hat{s} = 0.0877$

So 95% confidence intervals for 11 degrees of freedom are given by:

$$25.6 \pm t_{11} \times \frac{0.0877}{\sqrt{12}} = 25.6 \pm 2.201 \times \frac{0.0877}{\sqrt{12}}$$

$$= (25.54, 25.66)$$

Worked Example 11.2

The manufacturers state that the mean lifetime of a light bulb is supposed to be 100 hours. A sample of 6 bulbs gave lifetimes of 88, 96, 110, 120, 80, 94 hours respectively. Test the hypothesis at the 5% level of significance, that the bulbs match up to the manufacturers' statement.

Solution 11.2

H_0 : mean lifetime equals 100 hours
H_1 : mean lifetime not equal to 100 hours

Sample mean $= \frac{1}{6}[88 + 96 + 110 + 120 + 80 + 94] = 98$ hours

So $\bar{x} = 98$

$\hat{s}$ can be calculated from

$$\hat{s}^2 = \frac{\Sigma(x - \overline{x})^2}{(n - 1)} = \frac{1}{5}\left[10^2 + 2^2 + 12^2 + 22^2 + 18^2 + 4^2\right]$$

$$= 214.4$$

$$\Rightarrow \hat{s} = 14.64$$

$$t = \frac{\overline{x} - \mu}{\hat{s}/\sqrt{n}} = \frac{98 - 100}{14.64/\sqrt{6}} = -0.335$$

From the tables for t_5, we accept H_0 if $-2.571 \leqslant t \leqslant 2.571$; t lies in this range, hence we would be inclined to accept the manufacturer's claim.

Worked Example 11.3

A machine is thought to produce small lead weights with a mass of 25 g. To test the working of the machine, a sample of 12 weights was taken and found to have a mean of 27 g and standard deviation 2.2 g. Test the hypothesis that the machine is working properly using (i) 5% significance level; (ii) 1% significance level. Comment on the result.

Solution 11.3

Let $H_0 : \mu = 25, H_1 : \mu \neq 25$
A two-tailed test is appropriate.

$$\text{Here } \hat{s}^2 = \frac{n}{n - 1} s^2 = \frac{12}{11} \times 2.2^2 = 5.28$$

$$\Rightarrow \hat{s} = 2.298$$

$$t = \frac{27 - 25}{2.298/\sqrt{12}} = 3.015$$

(i) For 11 degrees of freedom, the 5% significance level is 2.201. Therefore we accept H_0 if $-2.201 \leqslant t \leqslant 2.201$. Since $t = 3.015$, we reject H_0 at the 5% level of significance.

(ii) For 11 degrees of freedom, the 1% significance level is 3.106. Therefore we accept H_0 if $-3.106 \leqslant t \leqslant 3.106$. Since $t = 3.015$, we accept H_0 at the 1% level of significance. Since H_0 is rejected at the 5% level of significance, but accepted at the 1% level of significance, it suggests that the machine may not be working properly.

11.6 Paired Samples

If two samples can be paired exactly, the differences between the values can be tested to see if they form a distribution with zero mean, assumed to be normal.

Worked Example 11.4

The times for 5 athletes over 200 m, on two different surfaces A and B are given in Table 11.1. Test the hypothesis that the performances are independent of the surface.

Table 11.1

Athlete	a	b	c	d	e
time on A	21.5	21.78	21.48	21.36	21.24
time on B	21.08	21.42	21.65	21.70	21.30

Solution 11.4

This will be a two-tailed test, since we are not interested which track is better. The table of differences is given in Table 11.2.

Table 11.2

Athlete	*Difference in time d*
a	0.42
b	0.36
c	−0.17
d	−0.34
e	−0.06

$H_0 : E(d) = 0$, i.e. no difference
$H_1 : E(d) \neq 0$
$\nu = 4$

$$\bar{d} = \frac{1}{5}[0.42 + 0.36 - 0.17 - 0.34 - 0.06] = 0.042$$

$$\hat{s}^2 = \{(0.42^2 + 0.36^2 + 0.17^2 + 0.34^2 + 0.06^2) - 5 \times 0.042^2\}/4$$
$$= 0.111 \qquad \text{so } \hat{s} = 0.334$$

$$\Rightarrow t = \frac{0.042 - 0}{0.334/\sqrt{5}} = 0.281$$

The two-tailed significance value at the 5% significance level for $\nu = 4$ is $t = 2.776$. Hence reject H_0 if $t > 2.776$. But $t = 0.281$, therefore performances on the two tracks appear to be independent of the surface.

Worked Example 11.5

Six chickens were housed in two types of accommodation, each for a period of 1 month. The numbers of eggs laid by each chicken in that time were as follows:

Chicken	*A*	*B*	*C*	*D*	*E*	*F*
			Eggs laid			
Accommodation 1	16	14	18	15	16	12
Accommodation 2	18	15	15	14	19	15

Test the hypothesis that the type of accommodation 2 is better.

Solution 11.5

If $d = x_2 - x_1$, then we have

$H_0 : E(d) = 0$
$H_1 : E(d) > 0$

A one-tailed test is appropriate here.
The values of d are 1, 2, −3, −1, 3, 3

so $\bar{d} = \frac{1}{6}[2 + 1 - 3 - 1 + 3 + 3] = 0.833$

and $\hat{s}^2 = \frac{1}{5}[(2 - 0.833)^2 + (1 - 0.833)^2 + (-3 - 0.833)^2 + (-1 - 0.833)^2 + (3 - 0.833)^2 + (3 - 0.833)^2]$

$= 5.767$

$$\Rightarrow \hat{s} = 2.40 \Rightarrow t = \frac{\bar{d} - \mu}{\hat{s}/\sqrt{n}} = \frac{0.833}{2.4/\sqrt{6}} = 0.85$$

The one-tailed significance value for $\nu = 5$ and 95% level is 2.015. Hence reject H_0 if $t > 2.015$. But $t = 0.85$. Hence the accommodation appears to make no difference.

Worked Example 11.6

A certain type of additive, MORMILE, in petrol is under investigation. Eight cars are tested with 10 gallons of petrol without the additive, and then re-tested with 10 gallons of petrol with the additive included. The results are given in Table 11.3.

Table 11.3

Car	*1*	*2*	*3*	*4*	*5*	*6*	*7*	*8*
Miles without additive	300	340	385	290	260	400	285	315
Miles with MORMILE	320	335	400	260	300	410	275	340

Test the hypothesis at the 5% level, that the MORMILE additive increases performance.

Solution 11.6

Since we are only interested in whether the performance level is improved, a one-tailed test is appropriate.

Find the difference d in the performances (Table 11.4). If we assume that the mean number of miles/10 gallons with the additive is μ_1 and the mean number of miles/10 gallons with MORMILE is μ_2, then $H_0 : \mu_1 = \mu_2$.

Now $E(d) = \mu_2 - \mu_1$

$H_0 : E(d) = 0$
and $H_1 : E(d) > 0$

$\nu = 7$
$\bar{d} = \frac{1}{8}(20 - 5 + 15 - 30 + 40 + 10 - 10 + 25) = 8.125$

$$\hat{s}^2 = \{(20^2 + 5^2 + 15^2 + 30^2 + 40^2 + 10^2 + 10^2 + 25^2) - 8 \times 8.125^2\}/7$$
$$= 492.4 \quad \Rightarrow \hat{s} = 22.19$$

So $t = \dfrac{8.125 - 0}{22.19/\sqrt{8}} = 1.036$

Table 11.4

Car	*Difference (d)*
1	20
2	−5
3	15
4	−30
5	40
6	10
7	−10
8	25

The one-tailed significance value at the 5% significance level for $\nu = 7$ is $t = 1.895$. Hence reject H_0 if $t > 1.895$. But $t = 1.036$. Therefore MORMILE does not appear to alter the performance.

11.7 Unpaired Samples

Consider two independent samples of size n_1 and n_2 respectively, with means $\bar{x}_1$ and $\bar{x}_2$.

Let H_0 : both samples are random samples from the same normal population of unknown mean μ and variance σ^2. We have that $\bar{x}_1$ is approximately $N\left(\mu_1, \dfrac{\sigma_1^2}{n_1}\right)$ and $\bar{x}_2$ is approximately $N\left(\mu_2, \dfrac{\sigma_2^2}{n_2}\right)$.

If we assume the samples are from the same population, then $\sigma_1 = \sigma_2 = \sigma$. It follows that $\bar{x}_2 - \bar{x}_1$ is approximately

$$N\left(\mu_1 - \mu_2, \left(\frac{\sigma^2}{n_1} + \frac{\sigma^2}{n_2}\right)\right)$$

We now define a statistic

$$t = \frac{(\bar{x}_1 - \bar{x}_2) - (\mu_1 - \mu_2)}{\hat{s}\sqrt{\dfrac{1}{n_1} + \dfrac{1}{n_2}}}$$

$$\hat{s}^2 = \frac{n_1 s_1^2 + n_2 s_2^2}{n_1 + n_2 - 2}$$ (see Worked Example 9.14 for various ways of calculating $\hat{s}^2$)

The number of degrees of freedom will be $n_1 + n_2 - 2$.

Worked Example 11.7

The average score of 30 pupils in a mathematics test was 65% with a standard deviation 18%. The average score of a different group of 20 pupils on the same test was 70% with a standard deviation of 12%. Is there any obvious difference between the performance of the two groups?

Solution 11.7

H_0 : performances are equal $\mu_1 = \mu_2$
$H_1 : \mu_1 \neq \mu_2$
$\bar{x}_1 = 65, s_1 = 18, \bar{x}_2 = 70, s_2 = 12, n_1 = 30, n_2 = 20$

$$\text{so } \hat{s}^2 = \frac{30 \times 18^2 + 20 \times 12^2}{30 + 20 - 2} = 262.5$$

$$\therefore \hat{s} = 16.2$$

$$t = \frac{(\bar{x}_1 - \bar{x}_2) - 0}{16.2\sqrt{\frac{1}{30} + \frac{1}{20}}} = \frac{65 - 70}{4.68} = -1.07$$

$\nu = 48$. The nearest value is $\nu = 60$, and the 95% value of t is 2.

So -1.07 is not significant.

There is no obvious difference in the performance.

*11.8 F-Ratio Test

In the previous section, the use of the t-test for unpaired samples depended on the assumption that the two samples had the same population variance. In order to test the validity of this assumption, the F-ratio test is used as follows:

For the two samples $\nu_1 = n_1 - 1$ and $\nu_2 = n_2 - 1$

$$\text{Let } F_{\nu_2}^{\nu_1} = \frac{\hat{s}_1^2}{\hat{s}_2^2}$$

H_0 : population variances equal
H_1 : one variance significantly greater than the other.

Hence if $\hat{s}_1^2$ is taken as the larger value of the variances a one-tailed test is appropriate.

If the two samples come from the same population, then clearly $F_{\nu_2}^{\nu_1}$ approaches unity as n_1 and n_2 become large.

Worked Example 11.8

Test whether the following two samples come from the same population.

Sample A	26	25	25	27	26					
Sample B	26	25	28	25	27	26	28	25	24	28

Solution 11.8

$\bar{x}_1 = 25.8, \bar{x}_2 = 26.2$

$\hat{s}_1^2 = \frac{1}{4}[0.2^2 + 0.8^2 + 0.8^2 + 1.2^2 + 0.2^2] = 0.7$

$$\hat{s}_2^2 = \tfrac{1}{9}[0.2^2 + 1.2^2 + 1.8^2 + 1.2^2 + 0.8^2 + 0.2^2 + 1.8^2 + 1.2^2 + 2.2^2 + 1.8^2]$$
$$= 2.178$$

Therefore $F_4^9 = \dfrac{2.178}{0.7} = 3.11$

From the tables, the 5% level for F_4^9 lies between 5.96 and 6.04. 3.11 is well inside this and suggests that the samples do come from the same population.

Worked Example 11.9

A random sample of size ten was chosen from London school-children of the same age. The heights (in cm) of the children chosen were:

							Total
Boys	142	125	127	122			516
Girls	130	122	117	125	123	121	738

On the assumption that the variance of boys' and girls' heights is the same, estimate the variance. Hence, assuming that the heights are normally distributed, find a 95% confidence interval for the difference in mean height of boys and girls of the age sampled.

What would be the result of a significance test of the null hypothesis that the mean height of boys and girls, of the chosen age, is equal? (OLE)

Solution 11.9

For the boys $\bar{x}_1 = \dfrac{516}{4} = 129$

$$s_1^2 = \tfrac{1}{4}[13^2 + 4^2 + 2^2 + 7^2] = 59.5$$

For the girls $\bar{x}_2 = \dfrac{738}{6} = 123$

$$s_2^2 = \tfrac{1}{6}[7^2 + 1^2 + 6^2 + 2^2 + 0^2 + 2^2] = 15.66$$

$$\Rightarrow \quad \hat{s}^2 = \frac{4 \times 59.5 + 6 \times 15.66}{4 + 6 - 2} = 41.5$$

For $\nu = 8$, the 95% confidence interval for the difference in the mean will be

$$(129 - 123) \pm t_8 \times \sqrt{41.5\left(\frac{1}{4} + \frac{1}{6}\right)}$$

$$= 6 \pm 2.3 \times 4.16 = (-3.57, 15.57)$$

Hence if mean height is assumed to be equal, the zero difference lies in this confidence interval. Hence H_0 is accepted.

11.9 Exercises

Exercise 11.1

(a) When an object is weighed on a chemical balance the readings obtained are subject to random errors which are known to be independent and normally distributed with mean zero and standard deviation 1 mg. A certain object is to be weighed 9 times on such a balance and the mean of the 9 readings is to be calculated. Find the probability that the mean of the 9 readings will be within 0.5 mg of the true weight of the object.

(b) Another weighing device is undergoing tests to determine its accuracy. A certain object of known true weight 50 mg was weighed 10 times on this device and the readings in mg were 49, 51, 49, 52, 49, 50, 52, 51, 49, 48.

(i) Calculate an unbiased estimate of the variance of the errors in readings using this device.

(ii) Calculate 95% confidence limits for the mean error in readings using this device.

Exercise 11.2

A random sample of receipts from a particular cashier in a large store showed totals of £8.50, £6.45, £3.40, £3.85, £10, £2.90, £1.73, £4.85, £4.60, £5.92, £8.66, £12.

(a) Find (i) the mean;
(ii) the standard deviation.

(b) Find 99% confidence limits for the mean.

(c) Test the hypothesis that the total takings per 100 receipts amount to £610 at the 0.05 level of significance.

Exercise 11.3

What is meant by the sampling distribution of a statistic? Show that the sampling distribution of the mean $\overline{x}$, calculated from a random sample $(x_1, x_2, \ldots, x_n)$ chosen from a normal distribution with mean μ and variance σ^2, also has mean μ. Find also the variance of the sampling distribution of the mean. Find central confidence intervals for the mean given the following data:

(i) A sample of 10 observations from a normal distribution: 5, 6, 1, 7, 10, 5, 12, 12, 14, 8. A 95% confidence interval is required.

(ii) A sample of 5 observations from a Poisson distribution: 14, 11, 13, 10, 12. A 99% approximate confidence interval, based on the normal approximation, is required.

(If you evaluate the variance or standard deviation of a set of data using a calculator, give a formula for the calculation.) (OLE)

Exercise 11.4

As a productivity deal, Ward Robe & Co. were prepared to pay a weekly bonus to assembly workers based on the number of cabinets assembled during the week. Prior to offering this deal, Sally Ree, a work-study engineer, selected a random sample of 10 assembly workers and timed them in assembling cabinets. From her figures, she calculated that the average number of cabinets assembled during a week is 45 units with a standard deviation of 8 units.

From these figures, what should the company set as the minimum number of cabinets to be completed in a week in order that at least 95% of the assembly workers should receive a productivity bonus. State any assumptions you have made.

Once the productivity deal had been accepted, Sally Ree chose again a random sample of 12 assembly workers and over a period of one week she found that the mean number of cabinets assembled was 50 units with a standard deviation of 6 units. Is there any evidence to suggest that the productivity deal has led to an increase in the number of cabinets assembled?

Exercise 11.5

To test a new chicken food additive, eight hens were given the normal food for three weeks and then were given the normal food together with the special additive for the next three weeks.

The number of eggs laid by each hen was as in Table 11.5. Perform a paired-sample t-test at the 5% level to investigate whether or not the additive results in a greater mean number of eggs. (SUJB)

Table 11.5

Hen	1	2	3	4	5	6	7	8
Fed with normal food, number of eggs	14	15	16	15	16	15	17	18
Fed with food + additive, number of eggs	15	16	16	16	17	17	18	18

Exercise 11.6

To test the effectiveness of a new powdered milk for babies a manufacturer used ten pairs of identical twins, each twin having the same birth weight as his brother or sister. Their diets were identical for a period of one month, except that one twin was fed with the old powder and the other with the new. The weights in kg after one month were as follows:

Twins	1	2	3	4	5	6	7	8	9	10
New Food	4.32	5.62	5.24	4.12	5.74	4.36	5.34	4.76	5.26	4.97
Old Food	4.30	5.43	5.26	4.18	5.41	4.22	5.39	4.61	5.14	4.86

Perform a paired-sample t-test to investigate whether or not the new food results in a greater mean weight than the old. What assumptions must be made in order for the t-test to be valid? Obtain 95% confidence limits for the difference in mean weights of twins fed on the two foods. (SUJB)

Exercise 11.7

A random sample of size n_1 is taken from a population P_1 whose mean is μ_1 and variance σ_1^2 and a random sample of size n_2 is taken from population P_2 with mean μ_2 and variance σ_2^2. Under what circumstances is it valid to test the hypothesis $\mu_1 - \mu_2 = 0$ using a two-sample t-test?

A machine fills bags of sugar and a random sample of 20 bags selected from a week's production yielded a mean weight of 499.1 g with standard deviation 0.63 g. A week later a sample of 25 bags yielded a mean weight of 500.2 g with standard deviation 0.48 g. Assuming that your stated conditions are satisfied perform a test to determine whether the mean has increased significantly during the second week. Test whether the mean during the second week could be 500 g. (Use a 5% significance level for both tests.) (SUJB)

Exercise 11.8

A random sample of 14 Midland banks produced a mean of 25.6 new accounts opened in a particular week, whereas another random sample of 10 Barclays banks produced a mean of 20.9 new accounts in the same week. The sample variances were 5.82 and 3.40 respectively.

(a) Find a 99% confidence interval for the mean of the number of new Midland bank accounts.
(b) Test the hypothesis that the means are equal at the 1% significance level.

Exercise 11.9

A manufacturer has speeded up one of the processes in his factory but is concerned that his action may have led to a loss of quality in the product. Eleven articles are chosen at random from the products of the current process and are assessed for quality on a scale from -2 to $+2$ ($0 =$ same standard as previous average quality, negative values represent a decrease in quality). The values, which may be taken to be approximately normally distributed, are:

$0, -1, 0, 1, -1, -2, -1, -1, 0, 1, -1$

(i) Carry out a test of whether there has been a deterioration in average quality.
(ii) Find a 95% confidence interval for the mean quality score of products from the new process.
(iii) Imagine you are the manufacturer. By considering (i) and (ii) state what you would conclude and what further action you might take. (OLE)

Exercise 11.10

The ability to withstand pain is known to vary from individual to individual. In a standard test a tiny electric shock is applied to the finger until a tingling sensation is felt. When this test was applied to a random sample of ten adults, the times recorded, in seconds, before they experienced a tingling sensation were

4.2, 4.5, 3.9, 4.4, 3.9, 4.5, 3.7, 4.8, 3.9, 4.2.

Use a single-sample t-test, at the 5% level, to test the hypothesis that the average time before an adult would experience a tingling sensation is 4.0 seconds. State any assumptions made.

It is believed that physical exercise increases the time until a tingling sensation is felt. The same ten adults were retested after performing a prescribed set of exercises. With the adults in the same order, the results, in seconds, were

4.4, 4.9, 4.0, 4.9, 3.8, 5.1, 3.6, 5.5, 3.8, 4.5.

Test this belief, at the 5% level, using a paired-sample t-test.

Explain why a paired-sample t-test is preferable to a two-sample t-test in this context. (UCLES)

12 Chi Squared Distribution

12.1 χ^2 Distribution

We have already seen how to fit a theoretical distribution to an observed distribution. We shall now define a distribution which will help us to measure how well theoretical data 'fits' the observed data.

The chi squared statistic, χ^2 (pronounced 'ki'), is defined as

$$\chi^2 = \sum \frac{(x_i - \bar{x})^2}{\sigma^2}$$

where each x_i is taken from a Normal distribution with a variance of σ^2. A sampling distribution for χ^2 for all samples of size n is called the *chi squared distribution.*

Like the t distribution, the χ^2 distribution is a family of density functions each one dependent on the number of degrees of freedom denoted by ν. The exact density function is far too complicated to quote and it is not necessary to know it when applying the distribution to statistical problems.

12.2 Properties

(i) Comparative shapes of the χ^2 distribution for different degrees of freedom are shown in Fig. 12.1.

(ii) The distribution is J shaped for $\nu = 1$ and $\nu = 2$, but as the number of degrees of freedom increases, the distribution becomes more symmetrical.

(iii) The distribution has a mean of ν and a variance of 2ν.

(iv) The statistic

$$\chi^2 = \frac{\Sigma(O - E)^2}{E}$$

where O is the observed frequency and E is the corresponding expected frequency of a particular event, can be approximated by the χ^2 distribution.

Once a value of χ^2 has been calculated, it is only necessary to determine the number of degrees of freedom in order to find the appropriate critical value for a stated level of significance.

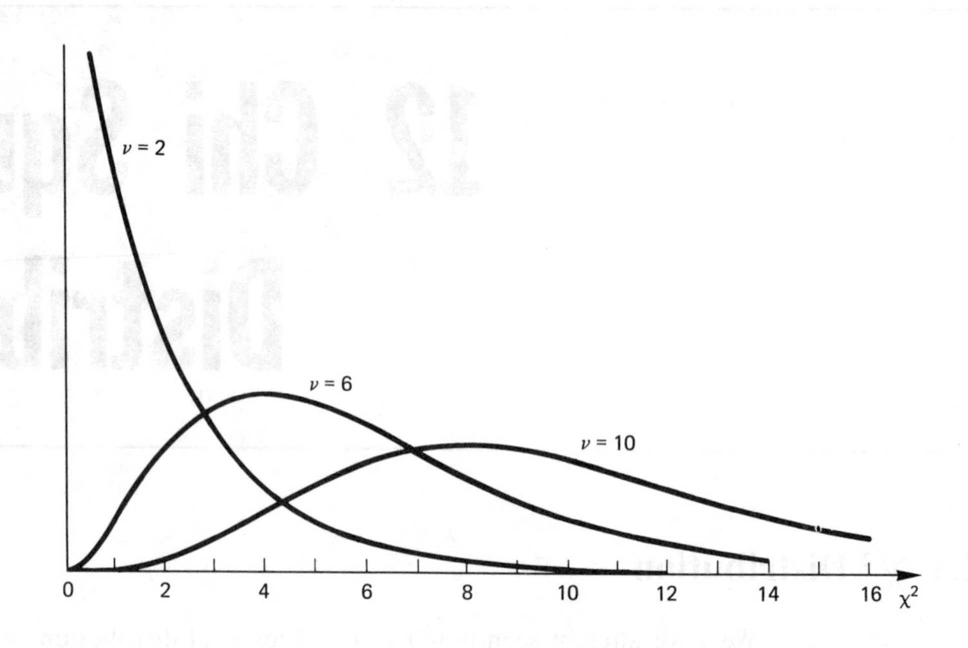

Figure 12.1

12.3 Use of the χ^2 Table

If, on the basis of a Null Hypothesis, the observed results are from an expected distribution, then intuitively the larger the differences between observed and expected frequencies, the more we are inclined to reject H_0. The χ^2 test is an acceptance or rejection of H_0 without the necessity of stating an Alternative Hypothesis. Although we may be suspicious that very small values of χ^2 are as a result of 'tampered' data, we rarely use χ^2 in a two-tailed way.

Worked Example 12.1

To test a hypothesis an experiment was carried out and resulted in a chi squared value of 18 with 9 degrees of freedom. Test at the 5% level of significance, whether or not the hypothesis can be rejected.

Solution 12.1

For a 5% level of significance, 9 degrees of freedom, the critical value of χ^2 is 16.919. As the observed value of χ^2 is greater than this we reject H_0 at the 5% level of significance.

12.4 Goodness of Fit

Worked Example 12.2

A die is thrown 180 times and the scores recorded as follows.

Score	1	2	3	4	5	6
Frequency	35	34	22	28	37	24

Is there any evidence that the die is biased?

Solution 12.2

The Null Hypothesis H_0, is that the die is fair and the expected frequencies are the same for each possible score. We shall test at the 5% level of significance.

In order to state the critical value we need to determine the number of degrees of freedom. To do this we have to see how many constraints we are applying to the expected frequencies.

In general:

number of degrees of freedom = number of classes − number of constraints

In this example $\Sigma E = \Sigma O$

so we have one constraint. Therefore the number of degrees of freedom, ν, is 5 and the critical value is 11.07. The value of χ^2 in Table 12.1 is not greater than the critical value so we have no reason to suspect that the die is biased.

Table 12.1

O	E	$O-E$	$(O-E)^2$	$(O-E)^2/E$
35	30	5	25	0.833
34	30	4	16	0.533
22	30	−8	64	2.133
28	30	−2	4	0.133
37	30	7	49	1.633
24	30	−6	36	1.200
				$\chi^2 = 6.465$

As we are using a continuous distribution as a model for the testing of discrete data, the result is meaningful provided two further conditions are satisfied:

(i) $\Sigma O \geqslant 50$
(ii) The minimum theoretical frequency for a class is 5.

Often, the tails of distributions give expected frequencies below 5 and in such instances we are entitled to group classes together.

Worked Example 12.3

A batch of seed was subjected to a new treatment. Following this treatment the seeds were sown under controlled conditions, five seeds per dish, 200 dishes. After one week the number of seeds germinating in each dish was counted and the results are shown in Table 12.2. Calculate the expected binomial probability distribution and the associated expected frequencies. Do the observed values fit the expected ones fairly well?

Table 12.2

Number germinating/dish	0	1	2	3	4	5
Frequency	4	11	23	54	75	33

Solution 12.3

The mean of the observed distribution is $(11 + 46 + 162 + 300 + 165)/200 = 3.42$.
For the expected binomial probability distribution, the probability of a seed germinating $= 3.42/5$
$= 0.684$

Number germinating/dish	*0*	*1*	*2*	*3*	*4*	*5*
Expected binomial probabilities	0.003	0.034	0.148	0.320	0.346	0.150
Expected frequencies (probabilities × 200)	0.6	6.8	29.5	63.9	69.2	30.0

The Null Hypothesis is that the observed values fit a binomial distribution with parameter 0.684. We shall test this at the 5% level of significance.

To determine the number of degrees of freedom, we should recognise that we shall have to combine the data from class 0 and class 1, giving only 5 classes in all. Also the number of constraints is now two:

(i) $\Sigma O = \Sigma E$, and
(ii) The mean of the two distributions must be the same.

So the number of degrees of freedom, $\nu = 5 - 2 = 3$
The critical value for a 5% level of significance is 7.815. The value of χ^2 in Table 12.3 is greater than the critical value and so we reject H_0 that the data fits a theoretical Binomial distribution. The result is highly significant ($< 1\%$).

Table 12.3

O	E	$O - E$	$(O - E)^2$	$(O - E)^2/E$
4, 11 (combined)	7.4	7.6	57.76	7.805
23	29.5	−6.5	42.25	1.432
54	63.9	−9.9	98.01	1.534
75	69.2	5.8	33.64	0.486
33	30.0	3.0	9.00	0.300
				$\chi^2 = 11.557$

12.5 Degrees of Freedom for Binomial, Poisson and Normal Distributions

For the three standard theoretical distributions, binomial, Poisson and Normal, the number of degrees of freedom for a given number of suitable classes, n, are given in Table 12.4. If population parameter(s) are unknown, they have to be calculated from the observed data.

Table 12.4

	Number of degrees of freedom	
Distribution	*Population parameter(s) known*	*Population parameter(s) unknown*
Binomial	$n - 1$	$n - 2$
Poisson	$n - 1$	$n - 2$
Normal	$n - 1$	$n - 3$

12.6 Confidence Interval for a Variance

For a random sample of size n drawn from a Normal population,

$$\chi^2 = \frac{\Sigma(x_i - \bar{x})^2}{\sigma^2}$$

$$\chi^2 = \frac{ns^2}{\sigma^2}$$

$$\text{Now } E\left(\frac{ns^2}{\sigma^2}\right) = \frac{n}{\sigma^2}\,.\,E(s^2) = \frac{n}{\sigma^2}.\frac{(n-1)}{n}\,\sigma^2$$

$$= n - 1$$

So $\dfrac{ns^2}{\sigma^2}$ has $(n - 1)$ degrees of freedom.

A 95% confidence interval therefore for the population variance is given by

$$\chi^2_{0.975} < \frac{ns^2}{\sigma^2} < \chi^2_{0.025}$$

$$\frac{ns^2}{\chi^2_{0.025}} < \sigma^2 < \frac{ns^2}{\chi^2_{0.975}}$$

with $(n - 1)$ degrees of freedom.

The corresponding interval for the standard deviation is

$$\frac{\sqrt{n}s}{\chi_{0.025}} < \sigma < \frac{\sqrt{n}s}{\chi_{0.975}}$$

Worked Example 12.4

The standard deviation of the journey time to work for a man over a period of 10 days is 5 minutes. What are the 95% confidence limits for the standard deviation of all journeys taken to work.

Solution 12.4

In order to make use of the intervals defined above we have to assume that the population is normally distributed.

For 9 degrees of freedom, the values of χ^2 such that the two tails give areas of 0.025 are 2.7 and 19.023.

The 95% confidence interval for the population standard deviation is

$$\frac{\sqrt{10}.5}{4.362} < \sigma < \frac{\sqrt{10}.5}{1.643}$$

$$\Rightarrow 3.62 < \sigma < 9.62$$

By increasing the size of the sample we could reduce the size of the interval.

12.7 Contingency Tables

Often it is necessary to investigate whether there is any significant difference between observed and expected frequencies when the distribution can be classified into two or more categories.

The mutually exclusive categories are tabulated in rows and columns and the result is called a *contingency table.*

Worked Example 12.5

Kinthatch, a roofing firm, carried out a survey to find out the political affiliation of its employees. Table 12.5 shows the results of the survey. Carry out a χ^2 test to determine whether there is any association between political affiliation and type of work.

Table 12.5

	Manual	*Non-manual*	*Total*
Conservative	22	16	38
Labour	53	8	61
Liberal/SDP	20	11	31
Total	95	35	130

Solution 12.5

On the basis that H_0 is that the classifications are independent, then a significant value of χ^2 gives evidence of a possible association between classifications. Under H_0 therefore, the expected frequencies are calculated as follows.

If the results are independent, then the proportion of manual to non-manual workers supporting each party should be 95/130 and 35/130, respectively. So the expected number who are affiliated to the Conservative party are

Manual	*Non-manual*
$\frac{95}{130} . 38$	$\frac{35}{130} . 38$

Continuing in this manner, the completed expected frequency contingency table is shown in Table 12.6.

Table 12.6

	Manual	*Non-manual*	*Total*
Conservative	27.8	10.2	38
Labour	44.6	16.4	61
Liberal/SDP	22.7	8.3	31
Total	95	35	130

Note: Expected values are given to one decimal place in this example.

In calculating the values for this table, two points emerge:

(i) A table entry can be found by multiplying together the marginal totals and dividing by the grand total.

(ii) Once two entries in the 'Manual' column have been calculated, the other entries can be determined from the marginal totals. The number of degrees of freedom therefore is 2.

In general:

> If a contingency table has r rows and c columns the number of degrees of freedom = $(r - 1) . (c - 1)$

At the 5% level of significance for 2 degrees of freedom the critical value is 5.991.

Table 12.7

O	E	$O - E$	$(O - E)^2$	$(O - E)^2/E$
22	27.8	−5.8	33.64	1.210
16	10.2	5.8	33.64	3.298
53	44.6	8.4	70.56	1.582
8	16.4	−8.4	70.56	4.302
20	22.7	−2.7	7.29	0.321
11	8.3	2.7	7.29	0.878
				χ^2 = 11.591

The result given in Table 12.7 is highly significant ($< 0.5\%$) and therefore we have evidence of association between the classifications.

12.8 Yates' Correction for Continuity

When the number of degrees of freedom is 1, we can improve the approximation to the continuous χ^2 distribution by using a continuity correction factor. This necessitates subtracting 0.5 from each the positive value (or 'modulus') $(O - E)$ before squaring.

Worked Example 12.6

Trubshire County Police produced a breakdown of statistics of serious car accidents in the county for the year 1985. The results were tabulated as in Table 12.8. Is there any evidence to suggest that motorists are safer wearing seat belts?

Table 12.8

	Wearing seat belt	*Not wearing seat belt*	*Total*
Killed	16	14	30
Not killed	314	56	370
Total	330	70	400

Solution 12.6

Under the Null Hypothesis that the classifications are independent, the expected frequency table is as shown in Table 12.9. At the 5% level of significance, 1 degree of freedom, the critical value is 3.841.

Table 12.9

	Wearing seat belt	*Not wearing seat belt*	*Total*
Killed	24.75	5.25	30
Not killed	305.25	64.75	370
Total	330	70	400

Table 12.10

O	E	$\lvert O-E\rvert-\frac{1}{2}$	$(\lvert O-E\rvert-\frac{1}{2})^2$	$(\lvert O-E\rvert-\frac{1}{2})^2/E$
16	24.75	–8.25	68.06	2.750
14	5.25	8.25	68.06	12.964
314	305.25	8.25	68.06	0.223
56	64.75	–8.25	68.06	1.051
				χ^2 = 16.988

The result in Table 12.10 is very highly significant (<0.1%) so there is strong evidence that motorists are safer wearing seat belts.

12.9 The Additive Property of χ^2

The χ^2 values from a number of repeated experiments can be added together and provided the corresponding degrees of freedom are also added, we can carry out a χ^2 test on the accumulated data.

Worked Example 12.7

Repeating an experiment three times, the χ^2 values were shown to be 5.2, 4.1 and 3.9 and each had 2 degrees of freedom. What conclusions can you draw?

Solution 12.7

Each of the individual experiments is not significant (the critical value for 2 degrees of freedom is 5.991). However if we use the additive property and combine the values, $\chi^2 = 5.2 + 4.1 + 3.9$
$= 13.2$ with 6 degrees of freedom.
The critical value for a 5% level of significance is 12.592, so a Null Hypothesis can be rejected from the combined data.

Worked Example 12.8

A supermarket manager predicts that the sales of 3 brands of washing powder A, B and C will sell in the ratio 1:2:3, respectively. The first 180 sales are distributed as in Table 12.11. Do the initial sales justify the manager's predictions?

Table 12.11

Brand	A	B	C	Total
Number sold	35	47	98	180

Solution 12.8

Under the Null Hypothesis that the expected ratios are 1:2:3 the expected frequencies are 30, 60 and 90. The number of degrees of freedom = 3 − 1 = 2. For a 5% level of significance, 2 degrees of freedom, the critical value is 5.991.

Table 12.12

O	E	$O-E$	$(O-E)^2$	$(O-E)^2/E$
35	30	5	25	0.833
47	60	−13	169	2.817
98	90	8	64	0.711
				χ^2 = 4.361

As $\chi^2 < 5.991$ the result (Table 12.12) is not significant and we have reason to believe that the manager's predictions are justified.

Worked Example 12.9

A random check of 200 tax discs on vehicles on the road in a certain area showed that 23 were invalid. Using a χ^2 test, is there any reason to believe that this proportion is significantly different from the national average of 8%? If a further independent sample of 250 vehicles from the same area showed that 28 had invalid tax discs, use the additive property of the χ^2 distribution to test whether the combined results are significant.

Solution 12.9

Using a χ^2 test we have only two observed values, those vehicles with tax discs and those without therefore we have only 1 degree of freedom and we shall have to use Yates' correction factor. The Null Hypothesis is that there is no difference between the observed proportion of invalid tax discs and the national average of 8%. At the 5% level of significance, one degree of freedom, the critical value is 3.841.

	O	E	$O-E-\frac{1}{2}$	$(O-E-\frac{1}{2})^2$	$(O-E-\frac{1}{2})^2/E$
Invalid tax disc	23	16	6.5	42.25	2.641
Valid tax disc	177	184	−7.5	56.25	0.306
					χ^2 = 2.947

As $\chi^2 < 3.841$, the result is not significant and we therefore accept the Null Hypothesis.

Repeating the analysis for the second sample:

	O	E	$O-E-\frac{1}{2}$	$(O-E-\frac{1}{2})^2$	$(O-E-\frac{1}{2})^2/E$
Invalid tax disc	28	20	7.5	56.25	2.813
Valid tax disc	222	230	−8.5	72.25	0.314
					χ^2 = 3.127

Again this result is not significant but by combining the 2 sets of data we can arrive at a value of χ^2 = 6.074 with 2 degrees of freedom. This result is now significant at the 5% level and we now have reason to believe that the area proportion of invalid tax discs is different than the national average.

Worked Example 12.10

One hundred electrical components are tested to see how many defects each has. The results are:

No. of defects	0	1	2	3	4	5	6	7 or more
No. of components	11	22	26	24	9	5	3	0

(i) Calculate the mean of the distribution.
(ii) Calculate the frequencies (to 1 decimal place) of the associated Poisson distribution having the same mean.
(iii) Perform a χ^2 goodness of fit test to determine whether or not the above results are likely to have come from a Poisson distribution, using a 5% significance level. (SUJB)

Solution 12.10

(i) $$\text{Mean} = \frac{0 + 22 + 52 + 72 + 36 + 25 + 18 + 0}{100}$$

$$= 2.25$$

(ii)

No. of defects (r)	*P(X = r)*	*Poisson frequencies*
0	0.1054	10.5
1	0.2371	23.7
2	0.2668	26.7
3	0.2001	20.0
4	0.1126	11.3
5	0.0506	5.1
6	0.0190	1.9
7 or more	0.0084	0.8

(iii) In testing the Null Hypothesis that the results are likely to have come from a Poisson distribution we have to determine the number of degrees of freedom.

Firstly we have imposed two restrictions on the theoretical distribution, namely the sum of the frequencies should be 100 and the mean of the distribution was estimated from the sample data.

Secondly, because some of the expected frequencies are smaller than 5 a number of the classes will have to be combined. By combining 5 or more defects we can carry out a valid χ^2 test.

The number of degrees of freedom = number of suitable classes – 2
= 4

For a 5% level of significance, the critical value is 9.488.

Table 12.13

No. of defects	O	E	$O-E$	$(O-E)^2$	$(O-E)^2/E$
0	11	10.5	0.5	0.25	0.024
1	22	23.7	−1.7	2.89	0.122
2	26	26.7	−0.7	0.49	0.018
3	24	20.0	4.0	16.00	0.800
4	9	11.3	−2.3	5.29	0.468
5 or more	8	7.8	0.2	0.04	0.005
					$\chi^2 = 1.437$

This value (Table 12.13) of χ^2 is certainly not significant and we therefore have reason to believe that the observed values come from a Poisson distribution with parameter 2.25.

Worked Example 12.11

Fergatron Electrics manufacture video cassette recorders at four different factories and the reliability of these machines is of great interest to the company. Table 12.14 relates to the performance of a number of similar machines during their first year of operation, together with the factory which had manufactured them.

Table 12.14

		Factory			
		A	B	C	D
Reliability	Repair needed	4	15	9	12
(first year)	Repair not needed	8	10	6	6

Use the χ^2 distribution and a 5% level of significance to test the hypothesis that there is no association between the factory of manufacture and the reliability of the product in the first year.

Subsequently Sam Pling, the company statistician, discovered that the above table was not actually the raw data, but that a junior assistant had divided all the original entries by 10 to reduce the arithmetic, and that this table was the result. How will the calculated value of your test statistic have to change to reflect this information, and what effect will this have on the result of the above test of association? Comment briefly upon these results.

(AEB 1983)

Solution 12.11

Under a Null Hypothesis that there is no association between the factory of manufacture and the reliability of the product in the first year, the number of degrees of freedom = (2 – 1).(4 – 1) = 3 and the critical value for a 5% level of significance is 7.815.

Using the marginal totals and the grand total to derive the expected frequencies we have,

		Factory				
		A	*B*	*C*	*D*	*Total*
Reliability (first year)	Repair needed	6.9	14.3	8.6	10.3	40
	Repair not needed	5.1	10.7	6.4	7.7	30
	Total	12	25	15	18	70

Calculating the value of χ^2 (Table 12.15), this value of χ^2 is less than the critical value so there is no reason to believe that there is any association between the factory of manufacture and the reliability of the product in the first year.

Table 12.15

O	E	$O-E$	$(O-E)^2$	$(O-E)^2/E$
4	6.9	−2.9	8.41	1.219
8	5.1	2.9	8.41	1.649
15	14.3	0.7	0.49	0.034
10	10.7	−0.7	0.49	0.046
9	8.6	0.4	0.16	0.019
6	6.4	−0.4	0.16	0.025
12	10.3	1.7	2.89	0.281
6	7.7	−1.7	2.89	0.375
				$\chi^2 =$ 3.648

If the actual observed values are 10 times greater than the stated values then

$$\chi^2 = \frac{\Sigma(10O - 10E)^2}{10E}$$

$$= \frac{100}{10}\,\frac{\Sigma(O-E)^2}{E}$$

$$= 10 \text{ times calculated chi squared statistic}$$

$$= 36.48$$

This value of χ^2 is very highly significant ($<0.1\%$) and we should have a very strong belief that an association exists.

Clearly in making the arithmetic easier by dividing through by a factor, the calculated value of χ^2 must be multiplied by the factor before any observations regarding its significance are made.

12.10 Exercises

Exercise 12.1

A random sample of 240 brown eyed individuals were classified according to the colour of their hair. The results are shown in the table below.

Hair colour	Red/Fair	Brown	Black
Observed	64	102	74

Is there any reason to suggest that having brown eyes is not related to hair colour?

Exercise 12.2

In the English Football League Divisions I and II, the number of goals scored in the 1982–83 season were as follows:

	Home	*Away*
Division I	810	454
Division II	750	494

Is there any difference in the proportions of goals scored Home and Away in the two Divisions?

Exercise 12.3

Prove that for a 2 x 2 contingency table

	x_1	x_2
y_1	a_1	a_2
y_2	a_3	a_4

$$\chi^2 = \frac{(a_1 + a_2 + a_3 + a_4).(a_1 a_4 - a_2 a_3)^2}{(a_1 + a_3)(a_2 + a_4)(a_1 + a_2)(a_3 + a_4)}$$

Exercise 12.4

A firm manufactures men's pyjamas in three sizes: small, medium and large. The firm's production is based on the assumption that the numbers of men requiring small, medium and large pyjamas will be on the ratio of 1:3:2. To test this assumption the firm measures its 60 male employees and the numbers requiring small, medium and large pyjamas are 5, 29, and 26, respectively. Use the χ^2 test to determine whether the results of this sample are consistent, at a 5% significance level, with the firm's assumption. (UCLES)

Exercise 12.5

A continuous random variable X has probability density function given by $f(x) = 0.2e^{-0.2x}$, $x \geqslant 0$.

Prove that $E(X) = 5$.

The survival times of 100 insects subjected to an insecticide are given in the table below.

Survival time in seconds	0–2	2–4	4–6	6–8	8–10	over 10
Number of insects	26	19	16	15	10	14

Use a χ^2 test, at the 5% level to determine whether the distribution of the random variable X defined above provides a good fit for the data. (UCLES)

Exercise 12.6

Under what circumstances would you expect a variate, X, to have a binomial distribution? What is the mean of X if it has a binomial distribution with parameters n and p.

A new fly spray is applied to 50 samples each of 5 flies and the number of living flies counted after one hour. The results were as follows:

Number living	0	1	2	3	4	5
Frequency	7	20	12	9	1	1

Calculate the mean number of living flies per sample and hence an estimate for p, the probability of a fly surviving the spray. Using your estimate calculate the expected frequencies (each correct to one place of decimals) corresponding to a binomial distribution and perform a χ^2 goodness-to-fit test using a 5% significance level. (SUJB)

Exercise 12.7

In processing grain in the brewing industry, the percentage extract recovered is measured. A particular brewery introduces a new source of grain and the percentage extract on eleven separate days is as follows:

95.2, 93.1, 93.5, 95.9, 94.0, 92.0, 94.4, 93.2, 95.5, 92.3, 95.4

Regarding the data as a random sample from a normal population, calculate a 90% confidence interval for the population variance. (AEB 1984)

Exercise 12.8

A firm uses three similar machines to produce a large number of components. On a particular day a random sample of 99 from the defective components produced on the early shift were traced back to the machines which produced them. The same was done with a random sample of 65 from the defectives produced on the late shift. The table shows the number of defectives found to be from each machine on each shift.

	Machine		
	A	*B*	*C*
Early shift	37	29	33
Late shift	13	16	36

(a) Use the χ^2 distribution to test, at the 5% significance level, the hypothesis that the probability of a defective coming from a particular machine is independent of the shift on which it was produced.

(b) Using the Early shift figures only, test, at the 5% significance level, the hypothesis that a defective is equally likely to have come from any machine.

(c) Comment briefly on your results given the following additional information.

All machines produce components at a similar rate.
During the early shift machines B and C were overhauled and so were in production for a much shorter time than A.

All machines were in production for the same time on the late shift.

(AEB 1984)

Exercise 12.9

Inflatia, in common with many other Western countries, is in the middle of an economic recession. As part of a nationwide enquiry into which economic measures will be most acceptable to the general public, a survey was undertaken in the town of Tucville. The responses to the question 'Would you support an incomes policy based on a flat rate increase of £500 per annum for every worker?' are summarised in Table 12.16, together with the employment status of the respondents.

Table 12.16

	Employment status			
Response	*Skilled and union member*	*Skilled and non-union member*	*Unskilled and union member*	*Unskilled and non-union member*
Yes	7	7	9	12
No	24	21	9	11
Don't know	29	27	17	27

Use the χ^2 distribution and a 5% level of significance to test the hypothesis that there is no association between response to the above question and employment status.

Form a new 2×2 contingency table from the above data by omitting all the 'Don't know' responses and then pooling the remaining responses to obtain one column for 'Skilled' and one column for 'Unskilled'. Use a 5% level of significance to test the hypothesis of no association between the factors in this new table. (AEB 1982)

Exercise 12.10

An experiment consists of 3 independent tosses of a coin, and the number of heads observed is i. The experiment is conducted independently $8n$ times, and on exactly r_i of these occasions the outcome of the experiment is i ($i = 0, 1, 2, 3$). If a chi squared test is conducted to investigate whether the coin is fair, write down expected values for each r_i, and hence show that the test statistic may be written

$$\frac{3r_0^2 + r_1^2 + r_2^2 + 3r_3^2}{3n} - 8n$$

The experiment is conducted 160 times, and the values of r_0, r_1, r_2, r_3 are respectively 15, 70, 50, 25. Do you consider the coin is a fair one? (OLE)

13 Correlation

13.1 Correlation and Scatter Diagrams

In section 4.3, we looked at the joint distribution of two random variables. A set of paired observations from two such variables is called a *bivariate distribution* and in this chapter we shall be interested in some measure of the relationship between the two variables. If a change in one variable is matched by a similar proportional change in the other variable, then the technique used to measure the degree of association is called *correlation.* It should be noted that although there may exist a high degree of correlation between two variables this does not imply a *causal* relationship between them. For instance one would expect a causal relationship to exist between the amount of rainfall in the summer months and the sale of umbrellas in the same period. However what might seem on the surface to be a causal relationship may in fact be spurious or nonsense correlation. Relationships can sometimes be shown to exist between the most obscure of variates. Justification is perhaps based only on the fact that both variates change with respect to time in a similar manner. Such an example is the number of television licences issued in a year and the corresponding number of juvenile delinquents.

Before quantifying the interdependence of two variables it is often useful to plot the paired observations on a *scatter diagram*. By inspection if a straight line can be drawn to fit the data reasonably well, then there exists a strong linear correlation between the variables.

Positive correlation: Both variables increase together.
Negative correlation: As one variable increases, the other decreases.

Examples of scatter diagrams are shown in Fig. 13.1.

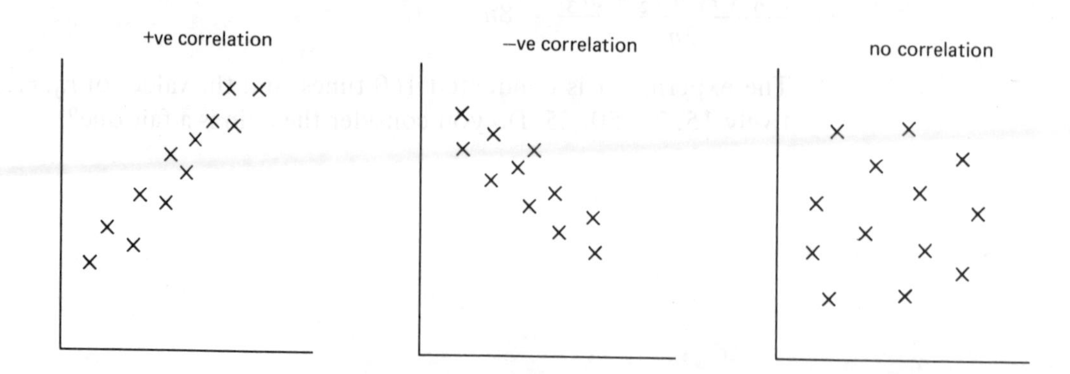

Figure 13.1

13.2 Product Moment Correlation Coefficient

The example in the diagram below shows some degree of positive correlation. In order to quantify this interdependence, we must first calculate $\Sigma(x_i - \bar{x})(y_i - \bar{y})$. Clearly positive contributions to this product can be found in the 1st and 3rd quadrant defined by the vertical line through the mean $\bar{x}$ and the horizontal line through the mean $\bar{y}$ (see Fig. 13.2).

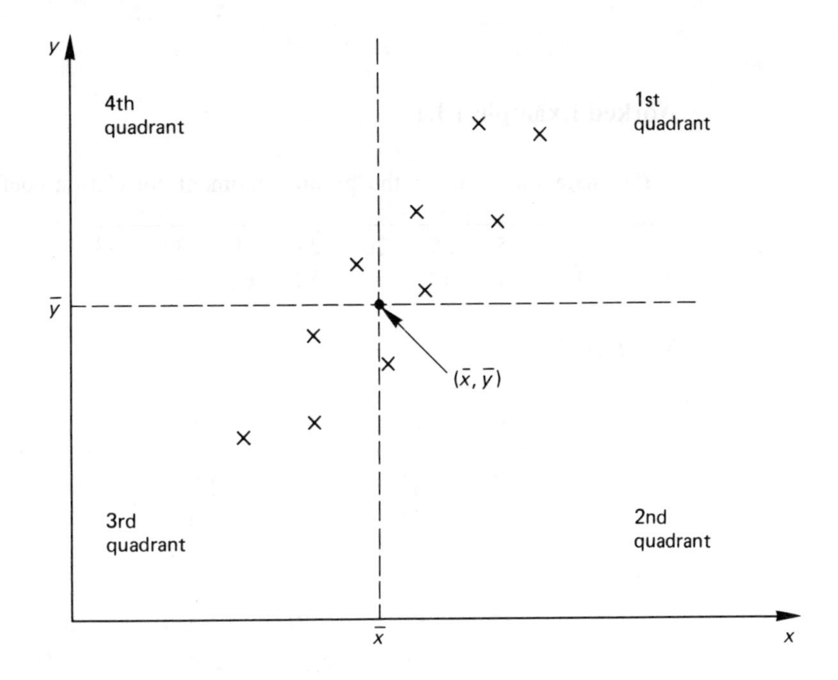

Figure 13.2

Negative contributions are obtained from the other two quadrants. So equal contributions from all quadrants leads to little or no correlation.

In order to arrive at a meaningful measure of correlation, two further factors must be taken into account regarding $\Sigma(x_i - \bar{x})(y_i - \bar{y})$.

(i) The size of $\Sigma(x_i - \bar{x})(y_i - \bar{y})$ is dependent on the number of paired observations, n.

(ii) The size is also dependent on the standard deviation of the x values and the standard deviation of the y values.

We therefore define the product moment correlation coefficient r, as

$$r = \frac{1}{n} \frac{\Sigma(x_i - \bar{x})(y_i - \bar{y})}{s_x \cdot s_y}$$

where s_x and s_y are the sample standard deviations of the two samples x and y. $\frac{1}{n}\Sigma(x_i - \bar{x})(y_i - \bar{y})$ is called the *covariance* of the sample and we shall represent this by s_{xy}.

Using $\bar{x} = \frac{\Sigma x_i}{n}$ and $\bar{y} = \frac{\Sigma y_i}{n}$ we can show that s_{xy} can be reduced to

$$s_{xy} = \frac{1}{n}\left(\Sigma x_i y_i - \frac{\Sigma x_i \Sigma y_i}{n}\right)$$

Using this alternative form for the numerator and similar alternatives for the denominator, we can arrive at a more useful form of the coefficient.

$$r = \frac{\Sigma x_i y_i - \dfrac{\Sigma x_i \Sigma y_i}{n}}{\sqrt{\left(\Sigma x_i^2 - \dfrac{(\Sigma x_i)^2}{n}\right)\left(\Sigma y_i^2 - \dfrac{(\Sigma y_i)^2}{n}\right)}}$$

Worked Example 13.1

Calculate the value of the product moment correlation coefficient for the following data.

x	23	25	25	26	27	29	32	33
y	45	51	49	54	52	60	58	58

Solution 13.1

	x	y	x^2	y^2	xy
	23	45	529	2025	1035
	25	51	625	2601	1275
	25	49	625	2401	1225
	26	54	676	2916	1404
	27	52	729	2704	1404
	29	60	841	3600	1740
	32	58	1024	3364	1856
	33	58	1089	3364	1914
TOTALS	220	427	6138	22 975	11 853

$$r = \frac{\Sigma x_i y_i - \dfrac{\Sigma x_i \Sigma y_i}{n}}{\sqrt{\left(\Sigma x_i^2 - \dfrac{(\Sigma x_i)^2}{n}\right)\left(\Sigma y_i^2 - \dfrac{(\Sigma y_i)^2}{n}\right)}}$$

$$= \frac{11\,853 - \dfrac{220\,.\,427}{8}}{\sqrt{\left(6138 - \dfrac{220^2}{8}\right)\left(22\,975 - \dfrac{427^2}{8}\right)}}$$

$= 0.869$

This can prove to be rather tedious but by suitably coding the data the arithmetic can be made easier and as we shall see, has no effect on the correlation coefficient.

Worked Example 13.2

Using the data from the previous example and defining $u = x - 27$ and $v = y - 52$, find the correlation coefficient between the variables u and v.

Solution 13.2

	u	v	u^2	v^2	uv
	-4	-7	16	49	28
	-2	-1	4	1	2
	-2	-3	4	9	6
	-1	2	1	4	-2
	0	0	0	0	0
	2	8	4	64	16
	5	6	25	36	30
	6	6	36	36	36
TOTALS	4	11	90	199	116

$$r = \frac{116 - \frac{4.11}{8}}{\sqrt{\left(90 - \frac{(4)^2}{8}\right)\left(199 - \frac{(11)^2}{8}\right)}}$$

$$= 0.869$$

Before we look at a test of significance we need to determine the range of possible values of r.

***Worked Example 13.3**

Show that the range of values of r is given by $-1 \leqslant r \leqslant 1$.

**Solution 13.3*

For any parameter λ, the expression $\Sigma((x_i - \bar{x}) + \lambda(y_i - \bar{y}))^2 \geqslant 0$
Expanding

$$\Sigma((x_i - \bar{x})^2 + \lambda^2 (y_i - \bar{y})^2 + 2\lambda (x_i - \bar{x})(y_i - \bar{y})) \geqslant 0$$

$$\Sigma(x_i - \bar{x})^2 + \lambda^2 \Sigma(y_i - \bar{y})^2 + 2\lambda\Sigma(x_i - \bar{x})(y_i - \bar{y}) \geqslant 0$$

Rearranging we have a quadratic in λ

$$\lambda^2 \Sigma(y_i - \bar{y})^2 + 2\lambda\Sigma(x_i - \bar{x})(y_i - \bar{y}) + \Sigma(x_i - \bar{x})^2 \geqslant 0$$

$$\Rightarrow 4(\Sigma(x_i - \bar{x})(y_i - \bar{y}))^2 \leqslant 4\Sigma(y_i - \bar{y})^2 \Sigma(x_i - \bar{x})^2$$

$$\Rightarrow \frac{1}{n^2}(\Sigma(x_i - \bar{x})(y_i - \bar{y}))^2 \leqslant \frac{\Sigma(y_i - \bar{y})^2}{n} \frac{\Sigma(x_i - \bar{x})^2}{n}$$

$$\Rightarrow s_{xy}^2 \leqslant s_x^2 \,.\, s_y^2$$

$$\Rightarrow \frac{s_{xy}^2}{s_x^2 s_y^2} \leqslant 1$$

$$\Rightarrow |r| \leqslant 1$$

So $-1 \leqslant r \leqslant 1$

13.3 The Significance of a Correlation Coefficient

The correlation coefficient r is an unbiased estimate of the population parameter ρ where $\rho = \dfrac{\text{Cov}(X, Y)}{\sigma_x \sigma_y}$ provided that a random sample is chosen. A large value of r suggests correlation but of course does not necessarily imply a causal relationship. If the Null Hypothesis H_0, is that there is no correlation between two variables, that is $\rho = 0$, then the distribution from which r comes is symmetrical and the t distribution is applied.

The test statistic $t = \sqrt{\dfrac{r^2(n-2)}{1-r^2}}$ is used with $n - 2$ degrees of freedom.

If we are testing the Null Hypothesis of a particular value of $\rho \neq 0$, then the distribution of r is far from symmetrical as Fig. 13.3 illustrates. In this instance, a

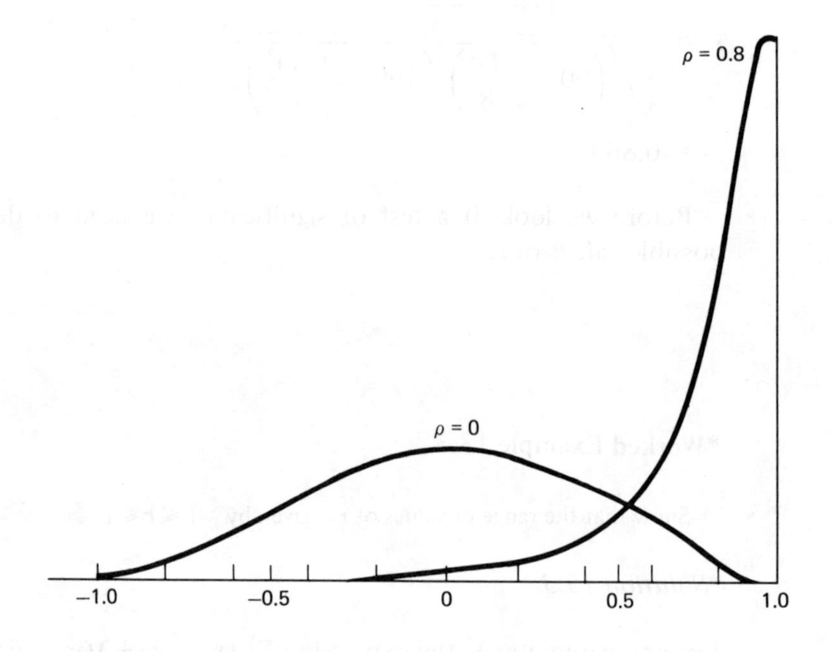

Figure 13.3

transformation is used to produce an approximately Normal distribution of the correlation coefficient. This is known as Fisher's transformation and is defined by

$$z_r = \frac{1}{2}\ln\left(\frac{1+r}{1-r}\right)$$

$$z_\rho = \frac{1}{2}\ln\left(\frac{1+\rho}{1-\rho}\right)$$

The distribution of $z_r - z_\rho$ is approximately Normal with mean 0 and standard error of $1/\sqrt{(n-3)}$.

To test the hypothesis, the test statistic z is calculated where

$$z = \frac{z_r - z_\rho}{1/\sqrt{(n-3)}}$$

The approximation is good for sample sizes of 50 or greater but it can be used for smaller sample sizes if we err towards caution in the testing procedure.

***Worked Example 13.4**

Test at the 5% level of significance the Null Hypothesis that there is no correlation between the data in Worked Example 13.1.

**Solution 13.4*

$H_0 : \rho = 0$
$H_1 : \rho \neq 0$

The computed value of r is 0.869 and the number of degrees of freedom is $(8 - 2) = 6$. The critical value for the required level of significance is therefore 2.447.

$$\text{The test statistic} \quad t = \sqrt{\frac{r^2\,(n-2)}{(1-r^2)}}$$

$$= \sqrt{\frac{0.869^2\,.\,6}{(1 - 0.869^2)}}$$

$$= 4.30$$

This is highly significant ($< 1\%$) and we reject H_0 and conclude that some correlation between the two sets of data exists.

***Worked Example 13.5**

A sample of 50 observations gives a correlation coefficient of 0.7. Is this observed value significantly different from a population correlation coefficient of 0.85?

**Solution 13.5*

$H_0 : \rho = .85$
$H_1 : \rho \neq .85$

Using the Fisher transformation

$$z_r = \frac{1}{2} \ln\left(\frac{1 + 0.7}{1 - 0.7}\right) = 0.867$$

$$\text{and} \quad z_\rho = \frac{1}{2} \ln\left(\frac{1 + 0.85}{1 - 0.85}\right) = 1.256$$

Under H_0, $z_r - z_\rho$ is approximately normally distributed with mean 0 and standard error of $1/\sqrt{47} = 0.146$. The critical values for a two-tailed test are $\pm$ 1.96.

$$\text{The test statistic} \quad z = \frac{0.867 - 1.256}{0.146}$$

$$= -2.66$$

This value is quite significant and we reject H_0 in favour of the Alternative Hypothesis.

13.4 Rank Correlation

If instead of the actual values, only the order in which items are ranked are given, then a correlation coefficient can be determined on the basis of the *ranks*. Below are described two such *rank correlation coefficients.*

13.5 Spearman's Coefficient of Rank Correlation

This is by far the most popular method of determining a coefficient of rank correlation and can be derived directly from the product moment coefficient using the ranks as the data.

Spearman's Rank Correlation Coefficient R, is defined as

$$R = 1 - \frac{6\Sigma d_i^2}{n(n^2 - 1)}$$

where $d_i = x_i - y_i$, the difference in rank value between observed pairs.

Worked Example 13.6

Determine Spearman's Rank Correlation Coefficient for the data of Worked Example 13.1.

Solution 13.6

You will notice that in both x and y we have tied ranks. To accommodate this minor complication we give each tied rank the mean of the positions occupied by the tied ranks. In this example the two 25's from the x distribution occupy positions 6 and 7 so both are given the tied rank of $6\frac{1}{2}$ (Table 13.1).

Table 13.1

x	8	$6\frac{1}{2}$	$6\frac{1}{2}$	5	4	3	2	1
y	8	6	7	4	5	1	$2\frac{1}{2}$	$2\frac{1}{2}$
d	0	$\frac{1}{2}$	$-\frac{1}{2}$	1	-1	2	$-\frac{1}{2}$	$-1\frac{1}{2}$
d^2	0	$\frac{1}{4}$	$\frac{1}{4}$	1	1	4	$\frac{1}{4}$	$2\frac{1}{4}$

$\Sigma d^2 = 9$

$$\text{So} \quad R = 1 - \frac{6\Sigma d_i^2}{n(n^2 - 1)}$$

$$= 1 - \frac{6 . 9}{8 . 63}$$

$$= 0.893$$

Compare this figure with the product moment correlation coefficient determined in Worked Example 13.1. The values differ slightly on two counts:

(i) there is a small error introduced in the assumption made about tied ranks;

(ii) the rank correlation coefficient makes no allowance for the difference in value of two items.

13.6 Kendall's Coefficient of Rank Correlation

This is not as popular as Spearman's as it is more awkward to calculate, however it is slightly more advantageous when testing for significance.

In order to determine the coefficient, it is necessary to consider all pairs of data from x and the corresponding data from y. If the pairs are in the same order, then 1 is added to the counter, P.

Kendall's Rank Correlation Coefficient τ, is defined as

$$\tau = \frac{2P}{\frac{1}{2}n(n-1)} - 1$$

Worked Example 13.7

10 drivers competing for a trophy completed two events and their finishing positions are shown below.

Driver	A	B	C	D	E	F	G	H	I	J
Event 1	9	2	3	6	4	1	10	8	5	7
Event 2	8	4	3	5	7	1	9	10	2	6

Calculate Kendall's rank correlation coefficient.

Solution 13.7

In order to determine the number of pairs, it is helpful to rank one set of data in order:

Driver	F	B	C	E	I	D	J	H	A	G
Event 1	1	2	3	4	5	6	7	8	9	10
Event 2	1	4	3	7	2	5	6	10	8	9

The data in Event 1 is in rank order and any pair of ranks from this Event are in the order 'low–high'. We have only to count up the number of corresponding pairs in Event 2 which are in the order 'low–high'. So for every pair of ranks in Event 2 in this order we add 1 to the pairs counter, P. As an example consider the following subset of the above data:

Driver	F	B	C
Event 1	1	2	3
Event 2	1	4	3

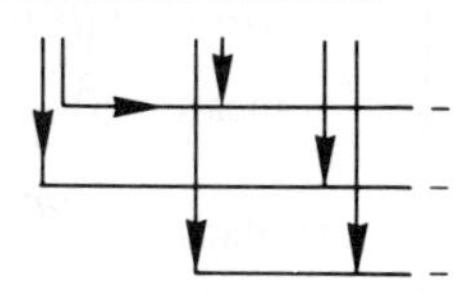

- 'low–high' so add 1 to pairs counter, P
- 'low–high' so add 1 to pairs counter, P
- 'high–low' so ignore

	Event 2	*Number of pairs*
To the right of:	1	9
	4	6
	3	6
	7	3
	2	5
	5	4
	6	3
	10	0
	8	1
	9	0

Total number of pairs, $P = 37$

$$\text{So} \quad \tau = \frac{2.37}{\frac{1}{2}.10.9} - 1$$

$$= 0.644$$

Tied ranks are dealt with by first averaging them, then if there is a tie in

(i) the second line of data, we add $\frac{1}{2}$ for each scoring pair;
(ii) in the first line, P is calculated twice. On the second calculation, the tied ranks are reversed and the mean score for the two values of P is calculated.

13.7 Significance of Rank Correlation Coefficients

In order to test the significance of either computed values, it is necessary to determine the probability that Σd^2 is equalled or exceeded in the case of Spearman's and the probability that P is equalled or exceeded in the case of Kendall's. This can prove to be a time consuming task and Table 13.2 gives numerical values of R and τ which are significant at the 5% and 1% level (one-tail) for sample sizes from 4 to 10 inclusive. If $n > 10$ the formulae are given for calculating the parameters of the approximating distribution.

Table 13.2

	Spearman's R		*Kendall's τ*	
	Significance level			
n	5%	1%	5%	1%
4	1.000		1.000	
5	0.900	1.000	0.800	1.000
6	0.829	0.943	0.733	0.867
7	0.714	0.893	0.619	0.810
8	0.643	0.833	0.571	0.714
9	0.600	0.783	0.500	0.667
10	0.564	0.746	0.467	0.600

	Spearman's R	*Kendall's τ*
$10 < n < 20$	*t-distributed* mean = 0 standard error = $\sqrt{\frac{1 - R^2}{n - 2}}$ $(n - 2)$ degrees of freedom	*Normally distributed* mean = 0 standard error = $\frac{4n + 10}{9n(n - 1)}$ Continuity correction factor: Reduce numerical value of τ by $\frac{1}{\frac{1}{2}n(n - 1)}$
$n > 20$	Normally distributed mean = 0 standard error = $\frac{1}{\sqrt{(n - 1)}}$	

Worked Example 13.8

A local vendor of ice cream in a seaside resort records his sales for ten consecutive days in August along with the maximum daily temperature.

Sales	(£)	x	82	79	66	59	50	57	61	83	85	52
Maximum temp.	(°C)	y	17	18	21	22	24	25	23	20	18	21

Calculate a coefficient of correlation and comment on your result.

Solution 13.8

Using the transformations: $u = x - 70, v = y - 20$

	u	v	u^2	v^2	uv
	12	−3	144	9	−36
	9	−2	81	4	−18
	−4	1	16	1	−4
	−11	2	121	4	−22
	−20	4	400	16	−80
	−13	5	169	25	−65
	−9	3	81	9	−27
	13	0	169	0	0
	15	−2	225	4	−30
	−18	1	324	1	−18
TOTALS	−26	9	1730	73	−300

$$r = \frac{(-300) - \dfrac{(-26).9}{10}}{\sqrt{\left(1730 - \dfrac{(-26)^2}{10}\right)\left(73 - \dfrac{9^2}{10}\right)}}$$

$$= -0.842$$

This high negative correlation suggests that there is an inverse relationship between daily temperature and ice cream sales which is not what we intuitively would expect. It is quite possible however that other indirect factors are affecting the results. For instance it is quite possible that the lower maximum temperatures coincide with a weekend. We would expect that the number of people at the resort to be higher at these times and consequently sales of ice cream could also be correspondingly higher.

***Worked Example 13.9**

During the summer of 1982 the National Leisure Council, on behalf of the Government, conducted a survey into all aspects of the nation's leisure time. Table 13.3 shows the amount spent per month on sporting pastimes and the total amount spent per month on all leisure activities for a random sample of thirteen young married men.

Table 13.3

Man		A	B	C	D	E	F	G
Amount on sport (£)	x	9.0	4.2	12.9	6.1	14.0	1.5	17.4
Total amount (£)	y	50.1	46.6	52.4	45.1	56.3	46.6	52.0
Man		H	I	J	K	L	M	
Amount on sport (£)	x	10.2	18.1	2.9	11.6	15.2	7.3	
Total amount (£)	y	48.7	56.0	48.0	54.1	53.3	51.7	

(a) Calculate the product-moment correlation coefficient for these data.

(b) Assuming that the above data comprise a random sample from a population with a bivariate normal distribution, test at the 1% level of significance the hypothesis $H_0 : \rho = 0$, where ρ is the true population correlation coefficient. (AEB 1983)

**Solution 13.9*

(a) Coding the data using $u = x - 10$ and $v = y - 50$ we have

	u	v	u^2	v^2	uv
	−1.0	0.1	1.00	0.01	−0.10
	−5.8	−3.4	33.64	11.56	19.72
	2.9	2.4	8.41	5.76	6.96
	−3.9	−4.9	15.21	24.01	19.11
	4.0	6.3	16.00	39.69	25.20
	−8.5	−3.4	72.25	11.56	28.90
	7.4	2.0	54.76	4.00	14.80
	0.2	−1.3	0.04	1.69	−0.26
	8.1	6.0	65.61	36.00	48.60
	−7.1	−2.0	50.41	4.00	14.20
	1.6	4.1	2.56	16.81	6.56
	5.2	3.3	27.04	10.89	17.16
	−2.7	1.7	7.29	2.89	−4.59
TOTALS	0.4	10.9	354.22	168.87	196.26

$$r = \frac{196.26 - \dfrac{(10.9)\,.\,(0.4)}{13}}{\sqrt{\left(354.22 - \dfrac{(0.4)^2}{13}\right)\left(168.87 - \dfrac{(10.9)^2}{13}\right)}}$$

$$= 0.824$$

(b) Null Hypothesis $H_0 : \rho = 0$
Alternative Hypothesis $H_1 : \rho \neq 0$

$$\text{Test statistic } t = \sqrt{\frac{(0.824)^2\ (13-2)}{1-(0.824)^2}}$$

$$= 4.82$$

For 11 degrees of freedom, this value is highly significant (< 1%) and therefore we reject H_0 in favour of H_1.

* Worked Example 13.10

The Government-financed Industrial Research Unit selected twelve factories at random from all those engaged in heavy engineering in this country. They were similar in all respects except size of workforce. Each factory was asked to indicate the percentage of employees absent from work for at least one whole day during a particular week. The results are tabulated in Table 13.4.

Table 13.4

Factory		*A*	*B*	*C*	*D*	*E*	*F*
Size of workforce (thousands)	x	1.1	1.9	3.0	4.2	5.1	5.8
Percentage of absentees	y	5	7	6	6	8	9
Factory		G	H	I	J	K	L
Size of workforce (thousands)	x	7.0	8.3	9.3	10.0	10.9	12.1
Percentage of absentees	y	11	12	14	20	43	60

(i) Calculate the product-moment correlation coefficient for these data.

(ii) Assuming that the above data comprise a random sample from a population with a bivariate normal distribution, test at the 5% level of significance the hypothesis $H_0 : \rho = 0.95$, where ρ is the true population correlation coefficient.

(iii) Calculate Spearman's rank correlation coefficient, and comment upon the value in the light of your previous results. (AEB 1982)

**Solution 13.10*

(i) Without coding it can be shown that

$$\Sigma x = 78.7$$
$$\Sigma y = 201$$
$$\Sigma x^2 = 660.71$$
$$\Sigma y^2 = 6601$$
and
$$\Sigma xy = 1856.5$$

$$\text{So } r = \frac{1856.5 - \dfrac{(78.7).(201)}{12}}{\sqrt{\left(660.71 - \dfrac{(78.7)^2}{12}\right)\left(6601 - \dfrac{(201)^2}{12}\right)}}$$

$$= 0.787$$

(ii) Null Hypothesis $H_0 : \rho = 0.95$
Alternative Hypothesis $H_1 : \rho \neq 0.95$
Using Fisher's Transformation,

$$z_r = \frac{1}{2}\ln\left(\frac{1+r}{1-r}\right) = \frac{1}{2}\ln\left(\frac{1.787}{0.213}\right) = 1.064$$

and
$$z_\rho = \frac{1}{2}\left(\ln\frac{1+\rho}{1-\rho}\right) = \frac{1}{2}\ln\left(\frac{1.95}{0.05}\right) = 1.832$$

The difference $z_r - z_\rho$ is approximately normally distributed with a mean of 0 and a standard error of $1/\sqrt{(n-3)}$.

Testing at the 5% level of significance, the critical values for a two-tailed test are $\pm$ 1.96.

Standardising, $z = \dfrac{(z_r - z_\rho) - 0}{\dfrac{1}{\sqrt{(n-3)}}}$

$$= \frac{1.064 - 1.832}{1/3} = -2.304$$

This result lies in the critical region and therefore we reject H_0 in favour of H_1.

(iii) Rank order of the data:

	A	B	C	D	E	F	G	H	I	J	K	L
x	1	2	3	4	5	6	7	8	9	10	11	12
y	1	4	$2\frac{1}{2}$	$2\frac{1}{2}$	5	6	7	8	9	10	11	12
d	0	-2	$\frac{1}{2}$	$1\frac{1}{2}$	0	0	0	0	0	0	0	0
d^2	0	4	0.25	2.25	0	0	0	0	0	0	0	0

$\Sigma d^2 = 6.5$

Spearman's rank correlation coefficient

$$R = 1 - \frac{6\Sigma d_i^2}{n(n^2 - 1)}$$

$$= 1 - \frac{39}{12.143}$$

$$= 0.977$$

There is a very strong rank correlation relationship but the significance of the product moment correlation coefficient suggests that the relationship is not linear.

Worked Example 13.11

The following are the grades (A to F) received by twelve candidates in Mathematics and French examinations, together with an overall grade based on all examinations taken.

Candidate	*1*	*2*	*3*	*4*	*5*	*6*	*7*	*8*	*9*	*10*	*11*	*12*
Overall	A	B	B	C	C	C	D	D	D	E	F	F
Mathematics	B	C	A	D	C	B	C	E	D	D	E	F
French	C	A	C	B	D	E	C	C	E	D	F	E

By calculating suitable correlation coefficients, decide which of Mathematics or French is the better predictor of the overall result. (SUJB)

Solution 13.11

Using Spearman's rank correlation coefficient

$$R = 1 - \frac{6\Sigma d_i^2}{n(n^2 - 1)}$$

The rankings are given in Table 13.5.

Table 13.5

Cand.	*Overall* (O)	*Maths* (M)	*French* (F)	$(O - M)$	$(O - M)^2$	$(O - F)$	$(O - F)^2$
1	1	$2\frac{1}{2}$	$4\frac{1}{2}$	−1.5	2.25	−3.5	12.25
2	$2\frac{1}{2}$	5	1	−2.5	6.25	1.5	2.25
3	$2\frac{1}{2}$	1	$4\frac{1}{2}$	1.5	2.25	−2.0	4.00
4	5	8	2	−3.0	9.00	3.0	9.00
5	5	5	$7\frac{1}{2}$	0	0	−2.5	6.25
6	5	$2\frac{1}{2}$	10	2.5	6.25	−5.0	25.00
7	8	5	$4\frac{1}{2}$	3.0	9.00	3.5	12.25
8	8	$10\frac{1}{2}$	$4\frac{1}{2}$	−2.5	6.25	3.5	12.25
9	8	8	10	0	0	−2.0	4.00
10	10	8	$7\frac{1}{2}$	2.0	4.00	2.5	6.25
11	$11\frac{1}{2}$	$10\frac{1}{2}$	12	1.0	1.00	−0.5	0.25
12	$11\frac{1}{2}$	12	10	−0.5	0.25	1.5	2.25
					46.50		96.00

Rank correlation between Overall and Mathematics

$$= 1 - \frac{6\Sigma(O - M)^2}{12.143}$$

$$= 1 - \frac{46.5}{286}$$

$$= 0.837$$

Rank correlation between Overall and French

$$= 1 - \frac{6\Sigma(O - F)^2}{12.143}$$

$$= 1 - \frac{96}{286}$$

$$= 0.664$$

Based on these figures we would put more reliability on the Mathematics results as a better overall predictor.

Worked Example 13.12

Different varieties of sweet peas are planted in plots whose distance from each other is measured, and the percentage of crossings of each variety is recorded in the table below. Calculate a rank correlation coefficient for the data and interpret your result.

Distance between plots (m)	32	60	60	84	103	112	121	135
Percentage of crossings of sweet peas	0.81	0.42	0.51	0.73	0.09	0.09	0.29	0.06

(L)

Solution 13.12

Calculating Spearman's rank correlation coefficient

Rankings:	Distance	1	$2\frac{1}{2}$	$2\frac{1}{2}$	4	5	6	7	8
	Percentage	1	4	3	2	$6\frac{1}{2}$	$6\frac{1}{2}$	5	8
	d	0	−1.5	−0.5	2	−1.5	−0.5	2	0
	d^2	0	2.25	0.25	4	2.25	0.25	4	0

So $\Sigma d^2 = 13$

$$R = 1 - \frac{6\Sigma d_i^2}{n(n^2 - 1)}$$

$$= 1 - \frac{6 . 13}{8 . 63}$$

$$= 0.845$$

From Table 13.2 this is significant at the 1% level.

If instead we had used Kendall's coefficient, then because of the tied rank in the upper line, two values of P have to be determined and the average of these two values found.

So, in the order shown,

$P_1 = 7 + 4 + 4 + 4 + 1\frac{1}{2} + 1 + 1 = 22\frac{1}{2}$

and reversing the tied ranks on the upper line,

$P_2 = 7 + 5 + 4 + 4 + 1\frac{1}{2} + 1 + 1 = 23\frac{1}{2}$

The average $P = (22\frac{1}{2} + 23\frac{1}{2})/2 = 23$

So Kendall's rank correlation coefficient

$$\tau = \frac{2P}{\frac{1}{2}n(n-1)} - 1$$

$$= \frac{46}{28} - 1$$

$$= 0.643$$

which is again highly significant.

13.8 Exercises

Exercise 13.1

For the following data, (a) plot the scatter diagram, (b) estimate the value of the product moment correlation coefficient r, (c) calculate the value of r.

x	−6	7	2	−1	4	−4	8	−2
y	9	−9	0	−6	−7	3	−10	4

Exercise 13.2

Watt Uhavin and Sam Egan were asked to judge the quality of ten drinks and place them in rank order. Watt Uhavin found the task relatively easy to place the drinks in order of merit, but Sam Egan found it difficult to separate the top three and decided to rank them equal first.

Drink	A	B	C	D	E	F	G	H	I	J
Watt Uhavin	10	8	4	1	2	6	7	5	9	3
Sam Egan	9	6	1=	1=	1=	8	7	5	10	4

Calculate:

(i) Spearman's rank correlation coefficient,
(ii) Kendall's rank correlation coefficient.

*Exercise 13.3

If X and Y are random variables and U and V are transformations of these variables such that $U = aX + b$ and $V = a'Y + b'$, show that $\text{Cov}(U, V) = aa'\text{Cov}(X, Y)$. Hence show that the product moment correlation coefficient of X and Y is the same as the product moment correlation coefficient of U and V.

Exercise 13.4

On the basis of the data in the table below, a psychologist asserts that, the older a child becomes, the fewer irrelevant actions he will make when performing a certain manual task.

Number of irrelevant actions made by children.

Age (years)	4	4	5	5	6	7	7	9	9	10	11
No. of irrelevant actions	10	13	12	9	10	7	12	8	7	5	6

(a) Draw a scatter diagram of the data.
(b) Find the product moment coefficient and, without further calculation, state whether you consider the psychologist's conclusion to be valid. (L)

Exercise 13.5

The ages, x years (given to one place of decimals), and heights, y cm (to the nearest cm), of 10 boys were as follows:

x	6.6	6.8	6.9	7.5	7.8	8.2	10.1	11.4	12.8	13.5
y	119	112	116	123	122	123	135	151	141	141

Given that $\Sigma x^2 = 899.80$, $\Sigma y^2 = 166\,091$ and $\Sigma xy = 12\,023.3$, calculate the linear correlation coefficient between x and y, and comment upon the result.
[Note: Linear c.c. = product moment c.c.] (UCLES)

*Exercise 13.6

In a workshop producing hand made goods a score is assigned to each finished item on the basis of its quality (the better the quality the higher the score). The number of items produced by each of 15 craftsmen on a particular day and their average quality score are given in Table 13.6.

Table 13.6

Craftsmen	*1*	*2*	*3*	*4*	*5*	*6*	*7*	*8*	*9*	*10*	*11*	*12*	*13*	*14*	*15*
Number of items produced, x	14	23	17	32	16	19	17	25	27	31	17	18	26	24	22
Average quality score, y	6.2	7.3	4.9	7.1	5.2	5.7	5.9	6.4	7.3	6.1	5.4	5.7	6.9	7.2	4.8

(a) Calculate the product moment correlation coefficient between x and y.
(b) Assuming the above data form a random sample from a bivariate normal distribution with correlation coefficient ρ, test the hypothesis $\rho = 0$, against the alternative $\rho \neq 0$, at the 5% significance level.
(c) The owner of the firm believes that the quality of the output is suffering because some craftsmen are working too quickly in order to increase bonus payments. Explain to him the meaning of your results, and state what evidence, if any, they provide for or against his belief. (AEB 1984)

Exercise 13.7

An educational psychologist obtained scores by 9 university entrants in 3 tests (A, B and C). The scores in tests A and B were as follows:

Entrant	*1*	*2*	*3*	*4*	*5*	*6*	*7*	*8*	*9*
A score	8	3	9	10	4	9	6	4	5
B score	7	8	5	9	10	6	3	4	7

Calculate a coefficient of rank correlation between these two sets of scores.

The coefficients obtained between the A and C scores was 0.71 and that between the B and C scores was 0.62. What advice would you give the psychologist if he wished to use less than three tests? (UCLES)

Exercise 13.8

A designer and a buyer were asked to forecast, in order of their sales potential, the ranks of ten styles of ladies' shoes for last Autumn's production. These forecasts and the ranks of the actual sales are given in Table 13.7. Calculate either Spearman's R or Kendall's τ to compare each forecast with the actual sales. Compare also the two forecasts with each other, commenting on the results. Using the data in the table suggest a method of making a forecast which might be more accurate than that of the designer or the buyer alone.

Table 13.7 Ranks of shoe styles according to actual sales and buyer's forecast and designer's forecast

Ranks	*Shoe styles*									
	A	*B*	*C*	*D*	*E*	*F*	*G*	*H*	*J*	*K*
Actual sales	5	8	1	9	4	2	7	10	6	3
Buyer's forecast	7	6	2	9	5	3	8	10	4	1
Designer's forecast	2	10	1	7	3	4	6	8	9	5

(L)

*Exercise 13.9

(a) The n pairs of observations (x_i, y_i), $i = 1, 2, \ldots, n$, are such that $x_1, x_2, \ldots, x_n$ are the integers from 1 to n in some order and $y_1, y_2, \ldots, y_n$ are also the integers from 1 to n in some order. Show that

$$\frac{1}{n}\sum_{i=1}^{n}(x_i - \bar{x})(y_i - \bar{y}) = \frac{1}{12}(n^2 - 1) - \frac{1}{2n}\sum_{i=1}^{n} d_i^2,$$

where

$$d_i = x_i - y_i, \bar{x} = \frac{1}{n}\sum_{i=1}^{n} x_i \text{ and } \bar{y} = \frac{1}{n}\sum_{i=1}^{n} y_i$$

(b) The table below gives the marks which nine candidates obtained in examinations in Mathematics and English.

Candidate number	*1*	*2*	*3*	*4*	*5*	*6*	*7*	*8*	*9*
Mark in Mathematics	52	86	25	33	85	55	57	54	46
Mark in English	40	65	48	57	50	41	39	63	34

Rank the data from top to bottom and calculate either the Spearman or the Kendall coefficient of rank correlation between the marks in Mathematics and English. Comment on the significance of your result. (UCLES)

Exercise 13.10

A clothing manufacturer collected the following data on the age, x months, and the maintenance cost, y (£), of his sewing machines.

Machine	*A*	*B*	*C*	*D*	*E*	*F*	*G*	*H*	*J*	*K*	*L*
Age, x	13	75	64	52	90	15	35	82	25	46	50
Maintenance cost, y	24	144	110	63	240	20	40	180	42	50	92

(a) Plot a scatter diagram of the data.
(b) Calculate the product moment correlation coefficient.
(c) Calculate Spearman's rank correlation coefficient.
(d) Without further calculation, estimate the product moment correlation coefficient and Spearman's rank correlation coefficient between X and $\log_e Y$ giving reasons for your estimates. (AEB 1987)

14 Linear Regression

14.1 Regression Analysis

By fitting a curve of best fit to the observations of a bivariate distribution, we can make estimates and predictions about the distribution of variables. Such an analysis is called *regression analysis* and lines of best fit are called *regression lines.*

14.2 Method of Least Squares

In calculating the line of best fit, the variable used as the predictor is assumed to have little or no error associated with it and is known as the *independent variable.*

The variable from which a variate is being predicted is subject to random errors and is known as the *dependent variable*.

For a bivariate distribution with X as the independent variable and Y the dependent variable, the equation of the line of best fit is $y = \alpha + \beta x$. However each y_i is subject to an error ϵ_i, and so each value of Y can be expressed by the equation

$$y_i = \alpha + \beta x_i + \epsilon_i$$

where

(i) α and β are unknown parameters,
(ii) the error terms, ϵ_i, are randomly distributed such that $E(\epsilon_i) = 0$,
(iii) each ϵ_i is independent of all other errors.

Estimates of α and β are made from the n pairs of observed data. If these estimates are a and b respectively, then

$$y_i = a + bx_i + \epsilon_i$$

The equation of the line of best fit is determined by the *method of least squares*, which requires that the sum of the squares of the vertical deviations from the line is a minimum (see Fig. 14.1).

We therefore require $\Sigma(y_i - a - bx_i)^2$ to be a minimum. For a minimum we use the methods of calculus (see Worked Example 14.6). The problem reduces to the following two equations:

$$\Sigma y_i = na + b\Sigma x_i$$

and $$\Sigma x_i y_i = a\Sigma x_i + b\Sigma x_i^2$$

These two equations are called the *Normal equations*, and can be solved simultaneously to find the values of a and b. The resultant equation, $y = a + bx$, is called the regression line of Y on X and b is the *regression coefficient.*

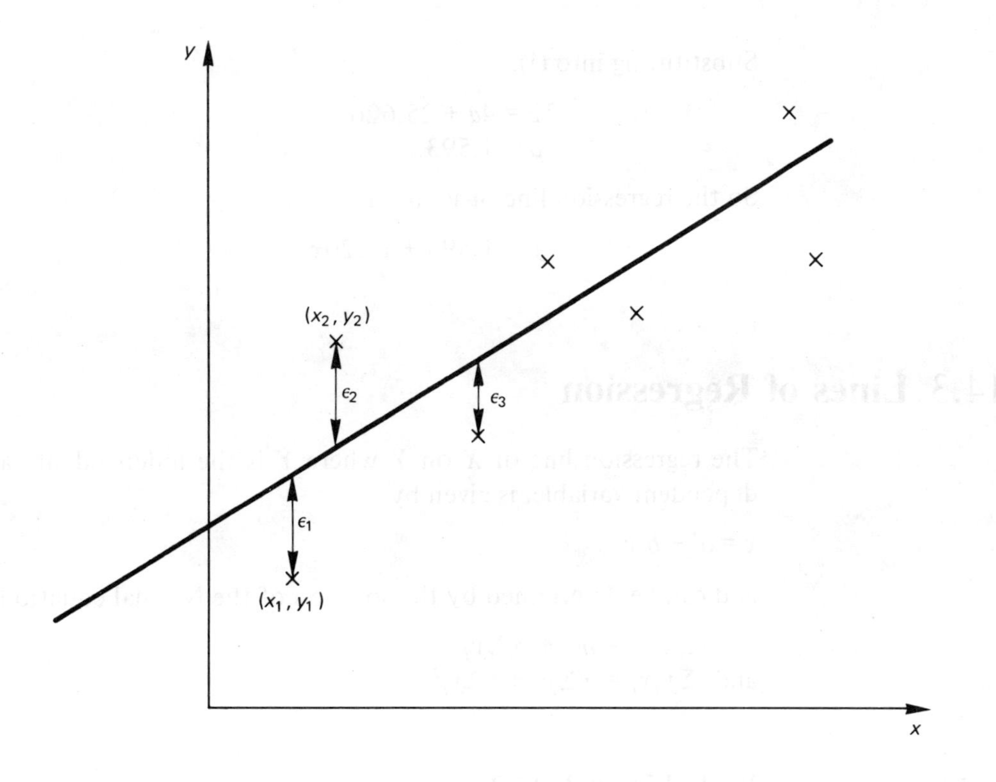

Figure 14.1

Worked Example 14.1

Determine the regression line of y on x for the following data.

x	3	4	6	8
y	4	9	7	12

Solution 14.1

	x	y	x^2	xy
	3	4	9	12
	4	9	16	36
	6	7	36	42
	8	12	64	96
Totals	21	32	125	186

Using the Normal equations:

$$32 = 4a + 21b \quad \ldots\ldots \text{(i)}$$
$$186 = 21a + 125b \quad \ldots\ldots \text{(ii)}$$

multiplying (i) by 21 and (ii) by 4

$$672 = 84a + 441b \quad \ldots\ldots \text{(iii)}$$
$$744 = 84a + 500b \quad \ldots\ldots \text{(iv)}$$

(iv) – (iii) $\quad 72 = 59b$

$$b = 1.2203$$

Substituting into (i),

$$32 = 4a + 25.626$$
$$a = 1.593$$

So the regression line of y on x is

$$y = 1.593 + 1.220x$$

14.3 Lines of Regression

The regression line of X on Y where Y is the independent variable, and X is the dependent variable, is given by

$x = a' + b'y$

and can be determined by the solution of the Normal equations

$$\Sigma x_i = na' + b'\Sigma y_i$$
and $$\Sigma y_i x_i = a'\Sigma y_i + b'\Sigma y_i^2$$

Worked Example 14.2

For the data of Worked Example 14.1, determine the regression line of X on Y.

Solution 14.2

We have already calculated the following:

$$\Sigma x_i = 21, \ \Sigma y_i = 32, \text{ and } \Sigma x_i y_i = 186$$

All we now require is $\Sigma y_i^2 = 4^2 + 9^2 + 7^2 + 12^2$
$= 290$

The two simultaneous equations are

$$21 = 4a' + 32b'$$
$$186 = 32a' + 290b'$$

Solving these gives

$$a' = 1.015$$
and $$b' = 0.529$$

So the regression line of X on Y is

$$x = 1.015 + 0.529y$$

See Fig. 14.2 for a graph of the two lines.

14.4 Properties

(i) $(\bar{x}, \bar{y})$ is the point of intersection of the two regression lines. By dividing the first Normal equation by n we have

$$\frac{\Sigma y_i}{n} = a + b\frac{\Sigma x_i}{n} \Rightarrow \bar{y} = a + b\bar{x}$$

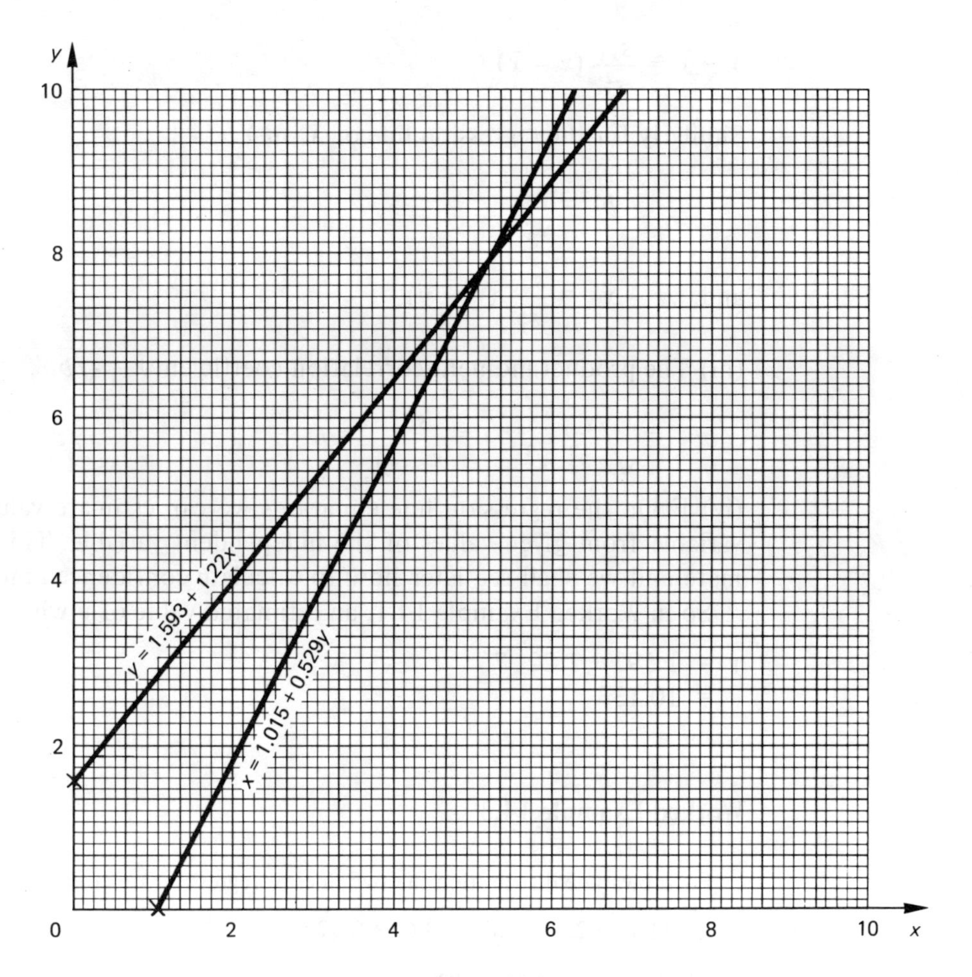

Figure 14.2

and $\quad \Sigma \frac{x_i}{n} = a' + b'\Sigma \frac{y_i}{n} \Rightarrow \overline{x} = a' + b'\overline{y}$

So $(\overline{x}, \overline{y})$ satisfies both regression equations.
In Fig. 14.2, the point of intersection is (5.25, 8).
(ii) For the regression line of Y on X

$$b = \frac{s_{xy}}{s_x^2}$$

Eliminating a from the Normal equations

$$\Sigma y_i \Sigma x_i = na\Sigma x_i + b\Sigma x_i \Sigma x_i \qquad \text{(i)}$$

and $\quad n\Sigma x_i y_i = na\Sigma x_i + nb\Sigma x_i^2 \qquad$ (ii)

(ii) – (i) $\; n\Sigma x_i y_i - \Sigma y_i \Sigma x_i = nb\Sigma x_i^2 - b(\Sigma x_i)^2$

Dividing throughout by n^2

$$\frac{1}{n}\left\{\Sigma x_i y_i - \frac{\Sigma x_i \, . \, \Sigma y_i}{n}\right\} = b\left\{\frac{\Sigma x_i^2}{n} - \left(\frac{\Sigma x_i}{n}\right)^2\right\}$$

$$\Rightarrow s_{xy} = bs_x^2$$

$$b = \frac{s_{xy}}{s_x^2}$$

Combining these two properties, we can arrive at an alternative form of the regression line of Y on X

$$y - \overline{y} = \frac{s_{xy}}{s_x^2}(x - \overline{x}) \qquad \text{(A)}$$

Similarly for the regression line of X on Y,

$$b' = \frac{s_{xy}}{s_y^2}$$

and $$x - \overline{x} = \frac{s_{xy}}{s_y^2}(y - \overline{y})$$

(iii) The product moment correlation coefficient $r = \sqrt{bb'}$.

$$\sqrt{bb'} = \sqrt{\frac{s_{xy} \cdot s_{xy}}{s_x^2 \cdot s_y^2}} = \frac{s_{xy}}{s_x \cdot s_y} = r$$

(iv) From the regression line equations we can estimate values of the dependent variable from given values of the independent variable. This will be a point estimate and we shall see later how to calculate confidence limits for such estimates.

So in Worked Example 14.1, an estimated value of y when $x = 7$ is given by

$$\begin{aligned} y &= 1.593 + 1.22 \,.\, 7 \\ &= 10.133 \end{aligned}$$

Worked Example 14.3

For a bivariate distribution the two lines of regression are given by

$y = 11.10 + 2.34x$ and $x = 3.67 + 0.37y$

Determine the value of the product moment correlation coefficient.

Solution 14.3

$$\begin{aligned} r &= \sqrt{\{\text{product of the regression coefficients}\}} \\ &= \sqrt{(b \,.\, b')} \\ &= \sqrt{(2.34 \,.\, 0.37)} \\ &= 0.930 \end{aligned}$$

Worked Example 14.4

In the table below, Y_i is the mass (in grammes) of potassium bromide which will dissolve in 100 grammes of water at a temperature of X_i°C.

X	10	20	30	40	50
Y	61	64	70	73	78

Find the equation of the regression line of Y on X. (SUJB)

Solution 14.4

The Normal equations for this data are

$$346 = 5a + 150b$$

and $$10\,810 = 150a + 5500b$$

Although we can solve these two simultaneous equations as they stand we can simplify the computations by coding the data in a similar manner to the method

used in the last chapter. The regression line is then determined directly from the alternative form ((A) page 212).
If the transformed variables U and V are defined by $U = c + dX$ and $V = c + dY$
then we can compute the regression line of V on U.

In this example we will let $U = \frac{1}{10}X - 3$

and $V = Y - 70$

	u	v	u^2	uv
	−2	−9	4	18
	−1	−6	1	6
	0	0	0	0
	1	3	1	3
	2	8	4	16
Totals	0	−4	10	43

$$\frac{s_{uv}}{s_u^2} = \frac{\Sigma u_i v_i - \dfrac{\Sigma u_i \Sigma v_i}{n}}{\Sigma u_i^2 - \dfrac{(\Sigma u_i)^2}{n}}$$

$$= \frac{43 - \dfrac{0 \times (-4)}{5}}{10 - \dfrac{0}{5}}$$

$$= 4.3$$

So the regression of V on U is given by

$$v + 0.8 = 4.3(u - 0)$$
$$v = 4.3u - 0.8$$

But we need the regression line of Y on X, so we substitute into this equation, the original transformations.

$$y - 70 = 4.3(0.1x - 3) - 0.8$$
$$y = 0.43x - 12.9 - 0.8 + 70$$
$$y = 0.43x + 56.3$$

14.5 Confidence Limits for Predicted Values

The observed values of the dependent variable are subject to independent random errors that are normally distributed with mean zero. In order to determine confidence limits we require either

(i) the standard deviation of the errors, σ_ϵ; or
(ii) some estimation of the standard deviation from the sample data.

If σ_ϵ is given then the confidence limits of a dependent variate is given by

$$\text{point estimate} \pm z\sigma_\epsilon$$

where z is determined from the level of confidence.

Worked Example 14.5

It is known that the observed values of y in Worked Example 14.1 are subject to independent random errors that are normally distributed with a mean of zero and a standard deviation of 0.7. Calculate the 95% confidence limits for the true value of y when $x = 7$.

Solution 14.5

We have already determined the point estimate of y as 10.13. So the 95% confidence limits for the true value

$= 10.13 \pm 1.96 \times 0.7$
$= 10.13 \pm 1.37$

If σ_ϵ is unknown, we need to estimate it from the sample data.

$$\text{Now } \Sigma\epsilon_i^2 = \Sigma(y_i - (a + bx_i))^2$$

$$= \Sigma y_i(y_i - (a + bx_i)) - \Sigma a(y_i - (a + bx_i)) - \Sigma bx_i(y_i - (a + bx_i))$$

$$= \Sigma y_i(y_i - (a + bx_i))$$

$$-a\Sigma(y_i - (a + bx_i)) \qquad (= 0 \text{ from Normal eqn (1)})$$

$$-b\Sigma(x_iy_i - ax_i - bx_i^2) \qquad (= 0 \text{ from Normal eqn (2)})$$

$$\text{So } \Sigma\epsilon_i^2 = \Sigma y_i^2 - a\Sigma y_i - b\Sigma x_iy_i$$

There are two restrictions in determining the squared deviations from the regression line, namely a and b. So the best estimate for σ_ϵ^2 is

$$\hat{s}_\epsilon^2 = \frac{\Sigma\epsilon_i^2}{n-2} = \frac{\Sigma y_i^2 - a\Sigma y_i - b\Sigma x_iy_i}{n-2}$$

The confidence limits of the predicted value of y are given by

$$a + bx \pm t_{(n-2)} \cdot \hat{s}_\epsilon$$

where $t_{(n-2)}$ is determined by the appropriate level of confidence and $n - 2$ degrees of freedom.

Worked Example 14.6

Show that by minimising $\Sigma(y_i - a - bx_i)^2$,

$$\Sigma y_i = na + b\Sigma x_i$$

$$\text{and } \Sigma x_iy_i = a\Sigma x_i + b\Sigma x_i^2$$

Solution 14.6

To minimise the two unknowns a and b in the expression we need to differentiate with respect to each variable in turn, keeping the other variable constant. This is a technique called *partial differentiation*. So differentiating with respect to a,

$$\frac{\partial}{\partial a}\Sigma(y_i - a - bx_i)^2 = -2\Sigma(y_i - a - bx_i)$$

For a minimum, $\Sigma(y_i - a - bx_i) = 0$

$$\Rightarrow \Sigma y_i = \Sigma(a + bx_i)$$

$$\Sigma y_i = na + b\Sigma x_i$$

and $\dfrac{\partial}{\partial b}\Sigma(y_i - a - bx_i)^2 = -2\Sigma x_i(y_i - a - bx_i)$

again for a minimum

$$\Sigma x_i(y_i - a - bx_i) = 0$$

$$\Sigma x_i y_i = \Sigma(ax_i + bx_i^2)$$

$$\Rightarrow \Sigma x_i y_i = a\Sigma x_i + b\Sigma x_i^2$$

Worked Example 14.7

As part of his research into the behaviour of the human memory, Hugo Nutts, a leading psychologist, asked 15 schoolgirls to talk for five minutes on 'my day at school'. Hugo then asked each girl to record how many times she thought that she had used the word nice during this period. Table 14.1 gives their replies together with the true values.

Table 14.1

Girl	*A*	*B*	*C*	*D*	*E*	*F*	*G*	*H*
True value x	12	20	1	8	0	12	12	17
Recorded number y	9	19	3	14	4	12	16	14
Girl	*I*	*J*	*K*	*L*	*M*	*N*	*O*	
True value x	6	5	24	23	10	18	16	
Recorded number y	5	9	20	16	11	17	19	

(i) Draw a scatter diagram of these data.
(ii) Fit a line by eye to these data, and label this line clearly.
(iii) Calculate the equation of the regression line of y on x and draw this line on your scatter diagram also. (AEB 1982)

Solution 14.7

(i) and (ii) See Fig. 14.3.
(iii) We shall use the Normal equations to find the equation of the regression line, but first we will code the data using

$$U = X - 10$$

and $$V = Y - 10$$

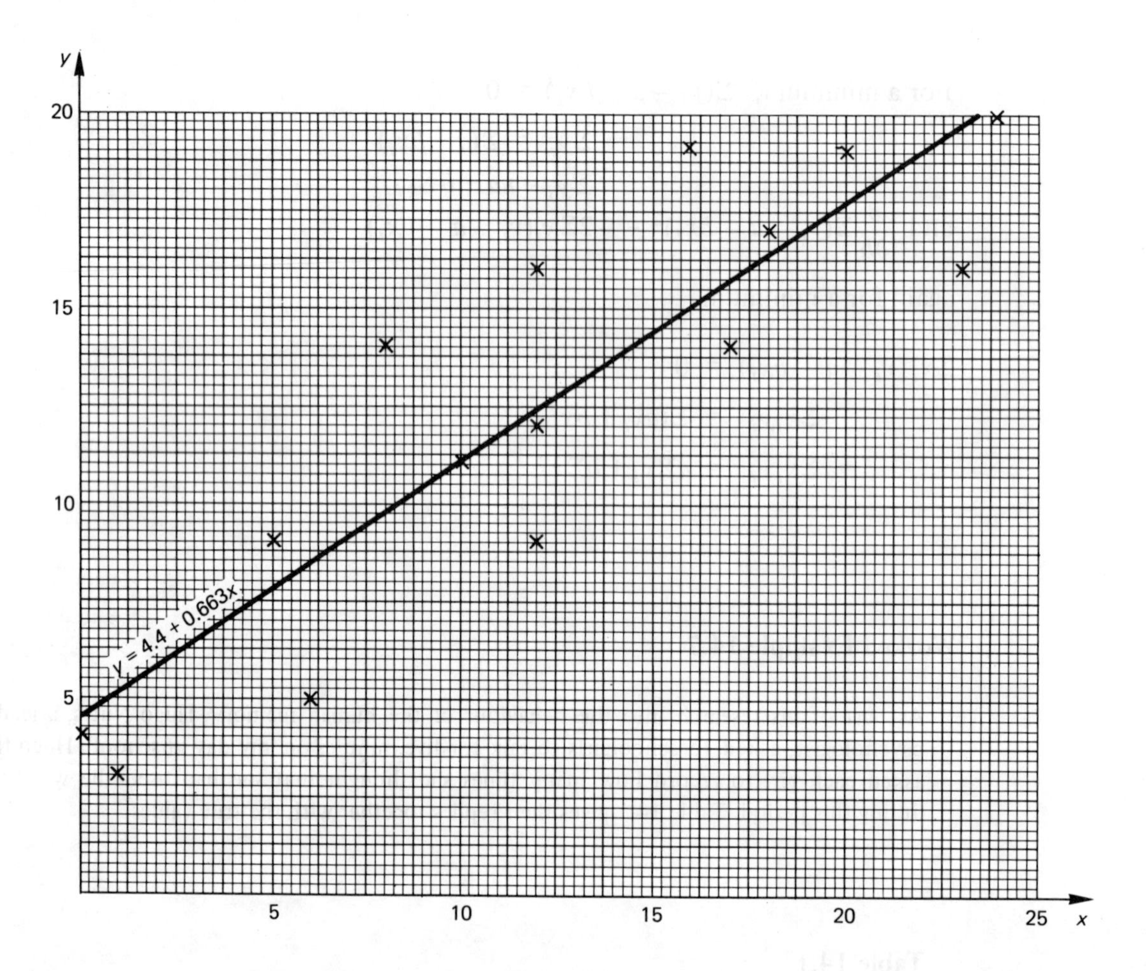

Figure 14.3

	u	v	u^2	uv
	2	-1	4	-2
	10	9	100	90
	-9	-7	81	63
	-2	4	4	-8
	-10	-6	100	60
	2	2	4	4
	2	6	4	12
	7	4	49	28
	-4	-5	16	20
	-5	-1	25	5
	14	10	196	140
	13	6	169	78
	0	1	0	0
	8	7	64	56
	6	9	36	54
TOTALS	34	38	852	600

$$\Sigma v_i = na + b\Sigma u_i$$

and $$\Sigma u_i v_i = a\Sigma u_i + b\Sigma u_i^2$$

$$38 = 15a + 34b \qquad \text{(i)}$$

$$600 = 34a + 852b$$

$$300 = 17a + 426b \qquad \text{(ii)}$$

multiplying (i) by 17 and (ii) by 15,

$$646 = 255a + 578b \qquad \text{(iii)}$$

$$4500 = 255a + 6390b \qquad \text{(iv)}$$

(iv) – (iii)

$$3854 = 5812b$$

$$b = 0.6631$$

and $$38 = 15a + 22.545$$

$$a = 1.0303$$

So the regression of V on U is

$$v = 1.0303 + 0.6631u$$

and by substitution the regression of Y on X is

$$y - 10 = 1.0303 + 0.6631(x - 10)$$

$$y = 11.0303 + 0.6631x - 6.631$$

$$\Rightarrow y = 4.4 + 0.663x$$

Worked Example 14.8

A spring hangs vertically downwards from a point P and a variety of weights T kg are attached to the other end of the spring. The extension, x, of the spring is measured in metres. It is known that T and x are connected by the formula $T = \lambda x + \epsilon$. Using the data in Table 14.2, estimate the equation of the regression line of x on T, and hence deduce estimates of λ and ϵ.

Explain why the regression line of x on T was used in preference to the regression line of T on x for estimating λ and ϵ. (L)

Table 14.2 The extensions (x) of a spring when weights (T) are applied to it

T (kg)	100	105	110	120	125	130
x (cm)	0.42	0.40	0.43	0.50	0.49	0.51

Solution 14.8

Coding the data, $U = \frac{1}{5}T - 23$

and $V = 100x - 45$

	u	v	u^2	uv
	−3	−3	9	9
	−2	−5	4	10
	−1	−2	1	2
	1	5	1	5
	2	4	4	8
	3	6	9	18
TOTALS	0	5	28	52

$$\frac{s_{uv}}{s_u^2} = \frac{\Sigma u_i v_i - \dfrac{\Sigma u_i \Sigma v_i}{n}}{\Sigma u_i^2 - \dfrac{(\Sigma u_i)^2}{n}}$$

$$= \frac{52 - \dfrac{0.5}{6}}{28 - \dfrac{0}{6}}$$

$$= 52/28$$

$$= 13/7$$

Now $\bar{u} = \Sigma u_i/n = 0$

and $\bar{v} = \Sigma v_i/n = 5/6$

So the regression line of V on U is given by

$$v - \frac{5}{6} = \frac{13}{7}u$$

The regression line of x on T is therefore

$$(100x - 45) - \frac{5}{6} = \frac{13}{7}\left(\frac{1}{5}T - 23\right)$$

$$100x = 0.37T + 3.12$$

$$x = 0.0037T + 0.0312$$

Making T the subject,

$$T = \frac{x - 0.0312}{0.0037}$$

$$\lambda = 1/0.0037 = 270$$

$$\text{and } \epsilon = -0.0312/0.0037 = -8.43$$

In determining the equation of a regression line there should be little or no error in the value of the independent variable. In this instance the weights T are deemed to be 'exact' whereas the extensions x are liable to measurement errors.

Worked Example 14.9

A bivariate distribution consists of n pairs (x_i, y_i). Given that $X_i = x_i + a$ and $Y_i = y_i + b$, where a and b are constants, prove that the regression coefficient of Y on X equals the regression coefficient of y on x.

Given that $n = 100$, $\bar{x} = -2.8$, $\bar{y} = 3.7$, $\Sigma(x + 2.8)^2 = 429$, $\Sigma(y - 3.7)^2 = 973$, $\Sigma(x + 2.8)(y - 3.7) = 865$, find the equation of the regression line of y on x and estimate the average value of y when $x = 5$. (L)

Solution 14.9

If $X_i = x_i + a \Rightarrow \bar{X} = \bar{x} + a$
and $Y_i = y_i + b \Rightarrow \bar{Y} = \bar{y} + b$

The regression coefficient of Y on X

$$= \frac{s_{XY}}{s_X^2}$$

$$= \frac{\Sigma(X_i - \bar{X})(Y_i - \bar{Y})}{\Sigma(X_i - \bar{X})^2}$$

$$= \frac{\Sigma((x_i + a) - (\bar{x} + a))((y_i + b) - (\bar{y} + b))}{\Sigma((x_i + a) - (\bar{x} + a))^2}$$

$$= \frac{\Sigma(x_i - \bar{x})(y_i - \bar{y})}{\Sigma(x_i - \bar{x})^2}$$

$$= \frac{s_{xy}}{s_x^2}$$

which is the regression coefficient of y on x.

For the second part let $X = x + 2.8 \Rightarrow \bar{X} = 0$
and $Y = y - 3.7 \Rightarrow \bar{Y} = 0$
and $\Sigma X^2 = 429$, $\Sigma XY = 865$.

The regression coefficient of Y on X is

$$b = \frac{\Sigma X_i Y_i - \dfrac{\Sigma X_i \Sigma Y_i}{n}}{\Sigma X_i^2 - \dfrac{(\Sigma X_i)^2}{n}} = \frac{\dfrac{\Sigma X_i Y_i}{n} - \overline{X}\,\overline{Y}}{\dfrac{\Sigma X_i^2}{n} - (\overline{X})^2}$$

$$= \frac{865/100}{429/100}$$

$$= 2.016$$

So the regression line of Y on X is given by

$$Y = 2.016X$$

and the regression line of y on x is

$$y - 3.7 = 2.016(x + 2.8)$$

$$y = 2.016x + 9.345$$

and when

$$x = 5, y = 19.43$$

Worked Example 14.10

A research worker, Dr Lin Guistic, gave each of eight children a list of 100 words of varying difficulty, and asked them to define the meaning of each one. Table 14.3 gives for each child, their age in years and the number of correctly defined words.

Table 14.3

Child		*A*	*B*	*C*	*D*	*E*	*F*	*G*	*H*
Age	x	2.5	3.1	4.3	5.0	5.9	7.1	8.1	9.4
Number of correct words	y	9	15	26	35	43	57	69	88

It is decided to fit a model of the form

$\log_e y = a + b \log_e x$

(a) Verify that this is a reasonable course to take by plotting the above data on log-log paper.
(b) Transform the data in the table, recording your values correct to one decimal place, to a form appropriate for fitting the above model.
(c) Calculate the equation of the regression line of $\log_e y$ on $\log_e x$ as given above.

(AEB 1983)

Solution 14.10

(a) This is left as an exercise for the reader. For a good model the plotted points should approximate to a straight line.

(b)

	$u = \log_e x$	$v = \log_e y$	u^2	uv
	0.9	2.2	0.81	1.98
	1.1	2.7	1.21	2.97
	1.5	3.3	2.25	4.95
	1.6	3.6	2.56	5.76
	1.8	3.8	3.24	6.84
	2.0	4.0	4.00	8.00
	2.1	4.2	4.41	8.82
	2.2	4.5	4.84	9.90
TOTALS	13.2	28.3	23.32	49.22

$\overline{u} = \Sigma u_i/n = 13.2/8 = 1.65$

$\overline{v} = \Sigma v_i/n = 28.3/8 = 3.5375$

$$\text{The regression coefficient of } V \text{ on } U = \frac{\Sigma u_i v_i - \dfrac{\Sigma u_i \Sigma v_i}{n}}{\Sigma u_i^2 - \dfrac{(\Sigma u_i)^2}{n}}$$

$$= \frac{49.22 - \dfrac{13.2 \,.\, 28.3}{8}}{23.32 - \dfrac{(13.2)^2}{8}}$$

$$= 1.64$$

The regression line of $\log_e y$ on $\log_e x$ is given by

$$\log_e y - 3.538 = 1.64 (\log_e x - 1.65)$$

$$\Rightarrow \log_e y = 1.64 \log_e x + 0.832$$

*Worked Example 14.11

Mark Aylot, a teacher at Liszt Squires High School assessed his pupils each year on two examination papers. In a class of 11 pupils, all sat the first test, but Corrie Lation, who achieved a mark of 28 on the first paper, was absent for the second paper. From the data of the rest of the class shown in Table 14.4, find an estimate for her mark on the second paper.

Stating any assumptions made, find the 95% confidence limits for the mark she would have obtained.

Table 14.4

Pupil	*A*	*B*	*C*	*D*	*E*	*F*	*G*	*H*	*I*	*J*
Paper 1 (x)	11	27	33	45	8	26	41	8	26	27
Paper 2 (y)	15	40	44	43	7	34	47	9	28	36

You may assume $\Sigma x_i = 252$, $\Sigma y_i = 303$, $\Sigma x_i^2 = 7854$

$\Sigma y_i^2 = 11185$, $\Sigma x_i y_i = 9271$

Solution 14.11

Using either the Normal equations or by calculating $(\bar{x}, \bar{y})$ and the regression coefficient we will arrive at

$$y = 2.891 + 1.088x$$

and when $x = 28$, $y = 33.36$

We shall assume that the deviations of the Paper 2 marks from the regression line are independently normally distributed with a mean of 0.

The variance of these deviations

$$\hat{s}_\epsilon^2 = \frac{\Sigma y_i^2 - a\Sigma y_i - b\Sigma x_i y_i}{n-2}$$

$$= \frac{(11185 - 2.891 \, . \, 303 - 1.088 \, . \, 9271)}{8}$$

$$= 27.77$$

So the standard error $\hat{s}_\epsilon = 5.27$ with 8 degrees of freedom.
The 95% confidence limits for Corrie Lation's mark on the second paper are

$$= 33.36 \pm t_8 \hat{s}_\epsilon$$
$$= 33.36 \pm 2.31 \times 5.27$$
$$= 33.36 \pm 12.17$$

14.6 Exercises

Exercise 14.1

Table 14.5 shows the relationship between the height of a shrub and the number of weeks after it was planted.

Table 14.5

Number of weeks, x	6	7	8	9	10	11
Height, y cm	105	113	124	136	142	151

Write down a coding formula to transform x to X in such a way that when $x = 6$, $X = -5$, and when $x = 11$, $X = 5$.

Use the method of least squares to fit a straight line relating y and X, taking X as the independent variable.

Use your coding formula to express y in terms of x. (OLE)

Exercise 14.2

The data in Table 14.6 relate to the price and engine capacity of new cars in January 1982.

Table 14.6

Car model	*A*	*B*	*C*	*D*	*E*	*F*	*G*	*H*	*J*	*K*
Price (£'s) (y)	3900	4200	5160	6980	6930	2190	2190	4160	3050	6150
Engine capacity (cc) (x)	1000	1270	1750	2230	1990	600	650	1500	1450	1650

(a) Plot a scatter diagram of the data.
(b) Calculate the regression line of y on x.
(c) Draw the regression line on the scatter diagram.

One particular customer regards low price and large engine capacity as the two most important factors in choosing a car. Examine your graph to suggest one model which, in January 1982, gave particularly good value for money, and three models you would have advised the customer not to buy. (AEB 1984)

Exercise 14.3

Under test, one year's wear on part of a car body can be simulated by one month's corrosion in the laboratory. The results in Table 14.7 show the number of years of simulated protection given by different thicknesses of paint. Find the equation of a suitable regression line. Estimate the thickness of paint needed for ten years' protection and comment on your result. (L)

Table 14.7

Thickness of paint (mm)	*0.25*	*0.30*	*0.35*	*0.40*	*0.45*
Number of years of protection	3.8	5.2	4.6	6.7	6.6

Exercise 14.4

In an agricultural experiment, a certain fertiliser is applied at different rates to ten identical plots of land, grass seed is sown, and several weeks later the mean height of the grass on each plot is measured. The results are shown in the table.

Rate of application of fertiliser (x g/m^2)	10	20	30	40	50	60	70	80	90	100
Mean height of grass (y cm)	9.8	11.0	13.2	11.6	14.8	14.8	16.0	19.4	20.6	18.8

Given that $\Sigma x_i = 550$, $\Sigma x_i^2 = 38\,500$, $\Sigma y_i = 150.0$, $\Sigma y_i^2 = 2374.08$, $\Sigma x_i y_i = 9212$, calculate,

(i) $\hat{s}_y^2$, the estimated population variance of y,
(ii) r, the estimated linear correlation coefficient between x and y.

Show that the equation of the estimated regression line of y on x is

$y = 0.117x + 8.59$

where the coefficients are given to three significant figures. (UCLES)

Exercise 14.5

Table 14.8 gives the mass of a baby elephant from 6 to 14 weeks after birth.

Table 14.8

Time (weeks) (t)	6	8	10	12	14
Mass (kg) (m)	103	106	112	115	120

The data are to be coded as $T = \frac{1}{2}(t - 10)$, $M = m - 100$. Plot these data on graph paper, and use the method of least squares to fit the following regression lines:

(a) regression line of M on T;
(b) regression line of m on t.

Use your regression lines to estimate the mass of the elephant after 16 weeks, and comment briefly on the reliability of this estimate. (OLE)

***Exercise 14.6**

(a) For a given bivariate distribution the equation of the line of regression of Y on X is $5y + 2x = 25$. The equation of the line of regression of X on Y is $2y + x = 11$. Find

(i) the means of X and Y;
(ii) the value of the linear (product moment) correlation coefficient between X and Y.

(b) The equation of the estimated line of regression of Y on X obtained from the five pairs of values (x_1, y_1), (x_2, y_2), (x_3, y_3), (x_4, y_4), (x_5, y_5), is $y = -0.55x - 0.05$. Given that $\sum_{i=1}^{4} x_i = 1$, $\sum_{i=1}^{4} y_i = -2$, $\sum_{i=1}^{4} x_i y_i = -10$, $\sum_{i=1}^{4} x_i^2 = 9$, find the values of x_5 and y_5. (UCLES)

***Exercise 14.7**

An experiment was conducted to determine the mass y grams of a given amount of chemical that dissolved in glycerine at x °C. The results of the experiment are given in the following table.

Temperature (x °C)	0	10	20	30	40	50
Mass (y g)	51.3	51.4	51.9	52.0	52.6	52.8

Assuming that the true value of y is linearly related to the value of x, obtain the least squares estimate of this relationship.

Assuming further that the temperatures used in the experiment were controlled accurately but that the measured values of y were subject to independent random errors which are normally distributed with mean zero and standard deviation 0.2 g, calculate 95% confidence intervals for

(i) the mass of chemical that will dissolve in glycerine at 0°C,
(ii) the additional mass of chemical that will dissolve in glycerine when the temperature is raised from 10°C to 20°C. (WJEC)

***Exercise 14.8**

A chemist set up an experiment to determine how a variable y varied with an associated variable x. In the experiment, x was set at the five values 0, 1, 2, 3, 4, respectively, and the corresponding values of y were observed. The chemist noted that the values of y increased fairly steadily with the increasing values of x, and, on applying the method of least squares to the results, the chemist produced the equation

$$y = 5.8 + 2.3x$$

(i) Find the value of $\overline{y}$, the mean of the five observed values of y. Suppose that the experimentally observed values of y are subject to independent random errors that are normally distributed with mean zero and standard deviation 1.1. Assuming that the true relationship connecting y and x is linear, calculate

(ii) a 95% confidence interval for the true value of y when $x = 4$,

(iii) a 90% confidence interval for the difference between the true values of y corresponding to $x = 1$ and $x = 4$, respectively. (WJEC)

Exercise 14.9

A small firm negotiates an annual pay rise with each of its twelve employees. In an attempt to simplify the process it is proposed that each employee should be given a score, x, based on his/her level of responsibility. The annual salary will be £$(a + bx)$ and the annual negotiations will only involve the values of a and b. The following table gives last year's salaries (which were generally accepted as fair) and the proposed scores.

Employee	A	B	C	D	E	F	G	H	I	J	K	L
x	10	55	46	27	17	12	85	64	36	40	30	37
Annual salary (£), y	5 750	17 300	14 750	8 200	6 350	6 150	18 800	14 850	9 900	11 000	9 150	10 400

You may assume that

$$(\Sigma x = 459 \quad \Sigma x^2 = 22\,889 \quad \Sigma y = 132\,600 \quad \Sigma xy = 6\,094\,750)$$

(a) Plot the data on a scatter diagram.

(b) Estimate values that could have been used for a and b last year by fitting the regression line $y = a + bx$ to the data. Draw the line on the scatter diagram.

(c) Comment on whether the suggested method is likely to prove reasonably satisfactory in practice.

(d) Without recalculating the regression line find the appropriate values of a and b if every employee were to receive a rise of
   (i) £500 a year
   (ii) 8%
   (iii) 4% plus £300 per year.

(e) Two employees, B and C, had to work away from home for a large part of the year. In the light of this additional information, suggest an improvement to the model. (AEB 1990)

*15 Further Topics

15.1 Non-parametric Tests

Most of the tests discussed in the previous chapters make assumptions about the distribution of a population and associated parameters. In particular, the Normal distribution is a frequent assumption which is necessary for the validity of such tests. One exception is the χ^2 distribution test of independence of data in cross-classifications. No assumptions are made about the distribution of the frequencies in the contingency table and it is known as a *non-parametric test.*

If little is known about the distribution of a population or that it is far removed from a Normal distribution, such tests can prove to be useful. However a word of caution. Because no restrictions are placed on non-parametric tests, the results are not quite as good as the standard methods when both can be applied.

15.2 Sign Test

An observation taken from any distribution is equally likely to be below the median as above it. In adopting the *sign test*, we note the sign of the differences between the observed values and the postulated population median. Under H_0, the distribution of + and – signs is binomially distributed and $P(+ \text{ sign}) = P(- \text{ sign}) = 1/2$.

The calculation of the binomial probabilities for a given sample size can determine the critical values for the tails of the distribution.

Worked Example 15.1

A population is thought to have a median of 12. The following data is a random sample from this population.

19 18 11 9 13 15 17 13

Solution 15.1

Null Hypothesis	H_0 : population median = 12
Alternative Hypothesis	H_1 : population median $\neq$ 12

Subtracting observed values from the postulated median and writing down the signs:

– – + + – – – –

Under H_0 the distribution of + and – signs is binomially distributed. If X is the distribution of + signs,

$$P(X \leqslant 2) = \left(\frac{1}{2}\right)^8 + 8\left(\frac{1}{2}\right)^8 + 28\left(\frac{1}{2}\right)^8$$

$$= 37\left(\frac{1}{2}\right)^8 = 0.145$$

For a 5% level of significance, two-tailed test, to reject H_0 we require the computed probability to be less than 0.025. This result is therefore not significant.

> **Extra note**: Any observations which are the same as the postulated median are ignored and this results in a reduction in the sample size.

In a similar manner we can carry out a sign test of paired comparisons. This involves subtracting the score of each subject in a given condition from the score of the matched pair in the other condition, noting only the sign of the difference and again testing the Null Hypothesis that the probability of a + sign equals the probability of a – sign or simply that the difference in population medians is zero.

Worked Example 15.2

10 brother–sister pairs were chosen at random. Each was given a test to assess reaction time. The results are tabulated as follows:

Brother's Reaction Time (secs)	5.2	4.6	5.6	4.5	4.8	5.4	4.3	5.2	4.9	4.7
Sister's Reaction Time (secs)	4.2	4.6	3.9	4.4	4.4	4.2	4.7	4.3	4.7	4.0

Carry out a sign test to see if there is any justification in the statement that on the whole, females have faster reaction times than males?

Solution 15.2

H_0 : males and females have similar reaction times.
H_1 : females have faster reaction times.

Subtracting matched pairs we have,

+ 0 + + + + – + + +

We ignore the results of the second pair and consider a sample of size 9. If X is the distribution of – signs then

$$P(X \leqslant 1) = \left(\frac{1}{2}\right)^9 + 9\left(\frac{1}{2}\right)^9 = \frac{10}{512} = 0.0195$$

For a one-tailed test this result is significant and we reject H_0 in favour of H_1.

15.3 Wilcoxon Signed Rank Test

The sign test makes no use of the magnitude of the differences between paired comparisons and as such is not as sensitive as other non-parametric tests. The *Wilcoxon signed rank test* of paired comparisons ranks the numerical differences in order of magnitude. Under a Null Hypothesis that there is no difference between the two sets of data, we would expect the sum of the positive ranks, T_+, to be close to the sum of the negative ranks, T_-. The Wilcoxon test therefore rejects H_0 if T_+ (or T_-) takes large or small enough values. Defining T as the minimum of T_+ and T_-, Table 16.5 gives the critical values for a 5% level of significance, one-tailed and two-tailed test. If T is less than or equal to the table value for the appropriate test, then we can reject H_0 at the 5% level of significance.

If the sample size > 15 then

$$E(T) = \frac{n(n+1)}{4}$$

and
$$\mathrm{Var}(T) = \frac{n(n+1)(2n+1)}{14}$$

and a Normal approximation with a continuity correction factor is used.

Worked Example 15.3

It has been claimed that the visual memory is more effective than the aural memory. To test this claim ten students were examined for visual and aural perception, the results being those shown in the table below. Set up appropriate null and alternative hypotheses and use the Wilcoxon signed rank statistic and a 5% significance level to test the null hypothesis.

Table 15.1 Memory test results (max. 20)

Visual	20	17	14	18	16	15	19	16	17	18
Aural	19	16	15	16	16	13	15	18	14	17

(L)

Solution 15.3

H_0 : no difference between visual memory and aural memory,
H_1 : visual memory is more effective than aural memory.

This is a one-tailed test and the differences are

1 1 −1 2 0 2 4 −2 3 1

(i) We are only concerned with +ve or −ve ranks so again zeros are ignored and the sample size is reduced to 9.
(ii) Tied ranks are averaged.

Differences in rank order according to magnitude:

	1	1	−1	2	0	2	4	−2	3	1
ranks:	$2\frac{1}{2}$	$2\frac{1}{2}$	$2\frac{1}{2}$	6		6	9	6	8	$2\frac{1}{2}$

So $T_- = 8\frac{1}{2}$

and $T_+ = 36\frac{1}{2}$

N.B. $T_+ + T_- = \dfrac{n(n+1)}{2}$

Taking the smaller of the two rank sums and calling it T, for a one-tailed test this value is not significant at the 5% level and therefore we accept H_0.

Although the Wilcoxon test is preferred to the sign test, the latter is quite useful when there is no quantification of data, i.e. differences are only 'better' or 'worse'.

15.4 Wilcoxon Rank Sum Test

The above tests cannot be carried out when either:

(i) there is no connection between corresponding items of data for samples of the same size,

(ii) when the two sample sizes are different.

In such instances we can use the *Wilcoxon Rank Sum test.* Suppose we have two sets of data of size n and m such that $n \leqslant m$. We collectively rank the two sets of data and sum the ranks of the smaller group calling the sum R. Under a Null Hypothesis that there is no significant difference between the two sets of data, all sets of n ranks are equally probable. As there are $\binom{m+n}{n}$ possible sets, the probability of each set is $1\Big/\binom{m+n}{n}$ and to calculate the probability that an observed value of R is significant – that is either too small or too large – can prove to be tedious. Table 16.6 shows critical values for a 5% level of significance, one- and two-tailed test, for sample sizes up to and including $n = 10$, $m = 10$. The value of R is significant if it is either less than or equal to the smaller table entry or greater than or equal to the larger table entry for the appropriate sample sizes and test.

For $n, m > 10$, the distribution of R is approximately Normal

$$E(R) = \frac{n(n+m+1)}{2}$$

$$\text{Var}(R) = \frac{nm(n+m+1)}{12}$$

Worked Example 15.4

Downsouth Bus Co. purchase a number of buses which they claim will provide a faster service than their old stock. Over a given route the times in minutes of 6 buses from the old stock and 8 of the new stock were recorded as follows:

Old Stock	22.1	23.9	20.9	23.6	25.0	24.2		
New Stock	19.8	22.1	20.3	20.8	21.2	22.3	21.0	21.5

Use a Wilcoxon Rank Sum test on this data to test the company's claim.

Solution 15.4

Ranking of data (Table 15.2). See next page.

Table 15.2

Data	*Rank*	*Data*	*Rank*
19.8	1	22.1	**$8\frac{1}{2}$**
20.3	2	22.1	$8\frac{1}{2}$
20.8	3	22.3	10
20.9	**4**	23.6	**11**
21.0	5	23.9	**12**
21.2	6	24.2	**13**
21.5	7	25.0	**14**

The ranks of the smaller set are underlined and the rank sum,

$$R = 4 + 8\tfrac{1}{2} + 11 + 12 + 13 + 14$$
$$= 62\tfrac{1}{2}$$

The claim by the company is that the new buses provide a faster service so the test is one-tailed and the table entries for $n = 6$, $m = 8$ and a one-tailed test are 31 and 59. As R exceeds 59 the result is significant at the 5% level.

15.5 Mann–Whitney *U* Test

This is very similar to the Wilcoxon Rank Sum test in that it allows for the testing of unmatched samples. In calculating the statistic U, we proceed as before in ranking the data from two samples of size n and m. The difference in this test is that the ranks of either sample are totalled and one of the two forms of U shown below are used.

Either $$U = nm + \frac{n(n+1)}{2} - R_n$$

or $$U = mn + \frac{m(m+1)}{2} - R_m$$

where R_n and R_m are the rank sums of the respective samples of size n and m.

Table 16.7 shows the critical values for a 5% level of significance, one- and two-tailed test. If the calculated value of U is less than or equal to the smaller table entry or greater than or equal to the larger table entry then the result is significant.

If n and m are sufficiently large (greater than 8) then U is approximately normally distributed.

$$E(U) = \frac{nm}{2}$$

and $$\text{Var}(U) = \frac{nm(n+m+1)}{12}$$

Worked Example 15.5

The pharmaceutical company, Everyit Haws & Co, decide to test whether or not a certain drug affects coordination in the recipient. 15 people are chosen at random, 9 of whom are administered the drug whilst the remaining 6 are administered a harmless placebo. The results of the coordination test are as follows:

Not receiving the drug	16	13	17	11	16	15			
Receiving the drug	16	12	8	10	14	7	9	12	13

Use a Mann-Whitney U test for the hypothesis that the drug has no effect on coordination.

Solution 15.5

Ranking the data (Table 15.3):

Table 15.3

Data	*Rank*	*Data*	*Rank*	*Data*	*Rank*
7	1	12	$6\frac{1}{2}$	15	11
8	2	12	$6\frac{1}{2}$	16	13
9	3	13	$8\frac{1}{2}$	16	13
10	4	13	$8\frac{1}{2}$	16	13
11	5	14	10	17	15

The underlined values are the ranks of the smaller set size and the sum of these values

$$= 5 + 8\tfrac{1}{2} + 11 + 13 + 13 + 15$$
$$= 65\tfrac{1}{2}$$

So $$U = 6 \cdot 9 + \frac{6(6+1)}{2} - 65\tfrac{1}{2}$$

$$= 9\tfrac{1}{2}$$

The hypothesis implies that we require a two-tailed test and the appropriate critical values for a 5% level of significance when $n = 6$ and $m = 9$, are 10 and 44. The calculated value of U is therefore just significant and further testing would be advisable.

Worked Example 15.6

Derive the sampling distribution of the statistic of the Mann-Whitney U test (or Wilcoxon Rank Sum test) on the Null Hypothesis of no difference between populations when both samples are of size three. State the critical region when the test is two-tailed, with significance level 10%.

A taxi company tests two types of tyres on the ten cars of its fleet. The length of life of the tyres, in thousands of miles, is:

Type A	29, 27, 23, 30
Type B	24, 37, 35, 19, 40, 31

Test if there is any difference in length of life of the two types of tyre. (OLE)

Solution 15.6

Using the Wilcoxon Rank Sum method there are $\binom{6}{3} = 20$ possible sets or combination of 3 ranks in a total of 6 ranks. The probability of set is therefore 1/20. The range of possible values of R is 6 to 15 inclusive and Table 15.4 gives the enumeration of possible rank scores.

Table 15.4

Ranks			*Score*	*Ranks*			*Score*
1	2	3	6	2	3	4	9
1	2	4	7	2	3	5	10
1	2	5	8	2	3	6	11
1	2	6	9	2	4	5	11
1	3	4	8	2	4	6	12
1	3	5	9	2	5	6	13
1	3	6	10	3	4	5	12
1	4	5	10	3	4	6	13
1	4	6	11	3	5	6	14
1	5	6	12	4	5	6	15

The sampling distribution of R is therefore

R	*6*	*7*	*8*	*9*	*10*
Probability	$\frac{1}{20}$	$\frac{1}{20}$	$\frac{2}{20}$	$\frac{3}{20}$	$\frac{3}{20}$
R	*11*	*12*	*13*	*14*	*15*
Probability	$\frac{3}{20}$	$\frac{3}{20}$	$\frac{2}{20}$	$\frac{1}{20}$	$\frac{1}{20}$

The critical region for a two-tailed test with a significance level of 10% is $R = 6$ or $R = 15$.

If instead we had considered the sampling distribution of the U statistic for the Mann–Whitney test, the distribution would have been identical to the one above but with values of U in the range 9 to 0 inclusive.

To test whether or not there is any difference in the length of life of the two types of tyres we shall use the Mann–Whitney U test. Ranking the data we have:

Data	19	23	24	27	29	30	31	35	37	40
Rank	1	*2*	3	*4*	*5*	*6*	7	8	9	10

The values *in italics* and underlined are the ranks of the smaller set size and the sum of these values is

$$2 + 4 + 5 + 6 = 17$$

So $$U = 4 \times 6 + \frac{4\,(4+1)}{2} - 17 \quad = 17$$

The hypothesis that there is no difference in length of life of the tyres implies that we require a two-tailed test and the appropriate critical values for a 5% level of significance when $n = 4$ and $m = 6$, are 2 and 22. The calculated value of U lies between these two values, therefore there is no evidence to suggest that there is any difference in length of life between the two types of tyres.

15.6 Analysis of Variance

When we wish to compare means of more than two populations, small samples having been drawn from these populations, the t-test is neither efficient nor proper. For an experiment with say seven sets of data there are 21 such tests and if each is tested at the 5% level of significance then in each test P(Type I error) = 0.05.

So P(Type I error overall) $= 1 - P$(No Type I error in 21 tests)

$$= 1 - (0.95)^{21} \quad = 0.66!!$$

For such tests we use the F-test of variances and the technique is known as the *analysis of variance*.

15.7 Single Factor Analysis of Variance

We shall first consider the situation where data is classified into groups on the basis of a single property. The Null Hypothesis is that the data is from a single population and any variation in the data is not due to the grouping of the data.

Putting this another way, the hypothesis reduces to the population means of the groups being identical.

To help in the explanation of the analysis we shall use the following example.

Worked Example 15.7

It is suspected that the output from three processing machines are different. In order to test this hypothesis, 5 random measurements for each machine are taken, each measurement being taken over the same interval of time. Carry out an analysis of variance on the data given in Table 15.5.

Table 15.5

	Machine		
	A	*B*	*C*
Output	11	9	7
	10	9	8
	8	9	7
	9	8	10
	12	10	8

Solution 15.7

Under the Null Hypothesis that the data is all from the same population, we can determine unbiased estimates of the population variance in three ways. Firstly, we will calculate column means and the overall or Grand mean.

A	*B*	*C*	*Overall mean*
$\bar{x}_1$	$\bar{x}_2$	$\bar{x}_3$	$\bar{x}$
10	9	8	9

(i) Regarding the 15 items of data as one sample we can calculate an estimate for the population variance.

The Total Sum of Squares of deviations from overall mean $\sum_{1}^{15} (x_i - 9)^2$

$= (12 - 9)^2 + (10 - 9)^2 + (8 - 9)^2 + \ldots + (8 - 9)^2$

$= 28$

and number of degrees of freedom = (15 − 1) = 14

Therefore an estimate for the population variance = 28/14 = 2

(ii) We can also estimate the population variance by pooling the three group variances together. To do this we need to calculate $\sum_{1}^{5} (x_{i1} - \bar{x}_1)^2$, $\sum_{1}^{5} (x_{i2} - \bar{x}_2)^2$ and $\sum_{1}^{5} (x_{i3} - \bar{x}_3)^2$ and combine this information in the pooled estimate,

$$\frac{\sum_{1}^{5}(x_{i1}-\overline{x}_1)^2+\sum_{1}^{5}(x_{i2}-\overline{x}_2)^2+\sum_{1}^{5}(x_{i3}-\overline{x}_3)^2}{n_1+n_2+n_3-3}$$

The numerator is known as *Within the Groups Sum of Squares* and the denominator, which in this instance is 12, is the number of degrees of freedom.

$$\text{Within the Groups Sum of Squares} = (12-10)^2+(10-10)^2+(8-10)^2 + \ldots + (10-8)^2+(8-8)^2 = 18$$

With a denominator of 12, an estimate of the variance = 18/12 = 1.5

(iii) Finally by replacing every item in a group by its group average we eliminate the variation within the group to leave only the variation between groups. So we have the data in Table 15.6.

Table 15.6

	Machine		
	A	*B*	*C*
	10	9	8
	10	9	8
Output	10	9	8
	10	9	8
	10	9	8

The sum of the squares of deviations from the overall mean = $\sum_{1}^{15}(x_i-9)^2$

$= 5(10-9)^2+5(9-9)^2+5(8-9)^2$

$= 10$

This is known as *Between the Groups Sum of Squares.*

The number of degrees of freedom is 2 as once two column entries are determined, the other 13 entries are dependent on these and the overall total of 135. (A simple way of remembering this is that it is always one less than the number of groups.)

So a third estimate of the variance = $\sum_{1}^{15}\frac{(x_i-9)^2}{2} = \frac{10}{2} = 5$

If H_0 is true, estimates (ii) and (iii) should be approximately equal in value and their ratio must approach 1. We therefore use the F-test with the appropriate degrees of freedom to test for significance. Before we compute the ratio it is worth summarising the calculated values in tabular form (see Table 15.7).

Table 15.7

	Sum of squares	*Degrees of freedom*	*Variance estimate*
Between Groups variation	10	2	5
Within Groups variation	18	12	1.5
Total variation	28	14	2

The Sum of Squares and Degrees of Freedom suggest an additive property and this can be confirmed using algebraic techniques.

Applying the F-ratio test $F^2_{12} = \dfrac{5}{1.5}$

$= 3.33$

The critical value for a 5% level of significance (from Table 16.3) is 3.89 and as this result is not exceeded by the calculated value, the result is not significant and we accept H_0.

In explaining the method, the solution to this problem has proved to be very lengthy and therefore the reader might find this summary useful.

1. In applying a One Factor Analysis of Variance we must assume:
   (i) under H_0, the data is taken from a single population;
   (ii) as F applies only to sampling from a Normal population, so the data should be normally distributed.
2. In calculating the table entries we make use of the additive property of the Sum of Squares and Degrees of Freedom by only computing two Sums of Squares. (Although the third could be calculated as a numerical check.)
   (i) Calculate the overall mean and the group means.
   (ii) Calculate the Total Sum of Squares of deviations from the overall mean

$$\bar{x} = \sum_1^n (x_i - \bar{x})^2$$

$$= \sum_1^n x_i^2 - n.\bar{x}^2$$

   with $n - 1$ degrees of freedom (n is the total sample size).
   (iii) Calculate the Between the Groups Sum of Squares of deviations from the overall mean $\bar{x}$ by replacing each group entry with its group average. The number of degrees of freedom = $c - 1$ where c is the number of groups.
3. On completion of the analysis of variance table, $F^{(c-1)}_{(n-c)}$ can be calculated and from the critical values in Table 16.3, it can be determined whether the result is significant.

Worked Example 15.8

In the previous example it is found that the measurement of 8 from machine A was in fact from machine C. Does this information affect our previous result?

Solution 15.8

The data now reads (Table 15.8)

Table 15.8

	Machine		
	A	*B*	*C*
Output	11	9	7
	10	9	8
	9	9	7
	12	8	10
		10	8
			8

(i) The group or column means are

A	B	C
10.5	9	8

and the overall or Grand mean is still 9.

(ii) The Total Sum of Squares of deviation from the overall mean remains the same as before and there are the same degrees of freedom.

(iii) Replacing every item of data in a group with its group average we have (Table 15.9)

Table 15.9

	Machine		
	A	B	C
	10.5	9	8
	10.5	9	8
Output	10.5	9	8
	10.5	9	8
		9	8
			8

The Between the Groups Sum of Squares is

$$4(10.5 - 9)^2 + 5(9 - 9)^2 + 6(8 - 9)^2$$

$$= 4 \times 2.25 + 0 + 6 \times 1$$

$$= 15$$

The number of degrees of freedom is one less than the number of groups $= 3 - 1 = 2$.

So the analysis of variance table is shown in Table 15.10, with the Within Groups data added.

Table 15.10

	Sum of squares	*Degrees of freedom*	*Variance estimate*
Between Groups variation	15	2	7.5
Within Groups variation	13	12	1.08
Total variation	28	14	

Applying the F-ratio test, $F^2_{12} = \dfrac{7.5}{1.08}$

$$= 6.94$$

The critical value for the appropriate degrees of freedom and a 5% level of significance is 3.89. So the result is significant and we reject H_0 that the data is all from the same population.

15.8 Two Factor Analysis of Variance

As an extension of the single factor analysis of variance, we can consider two variables of classification or groupings. As well as using the F-test to test any difference between column means, we can also test the row means of a second classification. This is known as *Two Factor Analysis of Variance*.

In the analysis of variance table for such tests the Between Groups variation is now replaced by Between Columns variation and Between Rows variation. The additive property for the Sum of Squares and degrees of freedom still applies and two F values are determined to test the hypotheses that the column means are the same and the row means are the same.

Worked Example 15.9

Instead of 5 random measurements for each machine in Worked Example 15.7, the table entries correspond to the outputs of 5 different individuals using each machine. Carry out a two factor analysis of variance on the data.

Solution 15.9

(i) To remind ourselves of the data we shall tabulate it along with column and row means (Table 15.11)

Table 15.11

		Machine			
		A	*B*	*C*	*Row means*
	1	11	9	7	9
Operator	2	10	9	8	9
output	3	8	9	7	8
	4	9	8	10	9
	5	12	10	8	10
Column means		10	9	8	

Again the overall mean is 9.

(ii) The Total Sum of Squares of deviation from the overall mean remains the same as before with the same degrees of freedom.

(iii) Likewise the Between the Columns Sum of Squares with associated degrees of freedom are the same as before. To calculate the Between the Rows Sum of Squares we eliminate the variation within the rows by replacing every item in a row with its row average (Table 15.12).

Table 15.12

		Machine		
		A	*B*	*C*
	1	9	9	9
Operator	2	9	9	9
output	3	8	8	8
	4	9	9	9
	5	10	10	10

The Between the Rows Sum of Squares is

$$3(9-9)^2 + 3(9-9)^2 + 3(8-9)^2 + 3(9-9)^2 + 3(10-9)^2$$
$$= 0 + 0 + 3 \times 1 + 0 + 3 \times 1$$
$$= 6$$

The number of degrees of freedom is one less than the number of rows

$$= 5 - 1 = 4$$

So the analysis of variance table is given in Table 15.13 with the addition of the variation left when between the rows and between the columns variations are both removed. It is often referred to as the residual.

Table 15.13

	Sum of squares	*Degrees of freedom*	*Variance estimate*
Between Columns variation	10	2	5
Between Rows variation	6	4	1.5
Residual	12	8	1.5
Total variation	28	14	

Under the assumption of the Null Hypothesis that there is no difference between machines and operators we compare in turn the estimates of the variance obtained from between rows and between columns with the estimate obtained when both column and row variations are removed (Residual).

The two values of F are 3.33 and 1 and for the appropriate degrees of freedom, neither of these are significant at the 5% level. The critical values for the two tests are 4.46 and 3.84 respectively.

15.9 Markov Chains

Consider the following:
In a given month if it rains on one day then the probability that it rains on the following day is 5/8, but if it is fine on one day, the probability that it is fine on the following day is 1/4.

Clearly there are a finite number of outcomes, or states, for the weather on any particular day and the sequence of outcomes for successive days is known as a *stochastic process*. In particular the weather on any one day is only dependent on the previous day's weather and such stochastic processes are known as *Markov chains.*

15.10 Transition Matrices

The probability of a particular outcome or state on a given day which is dependent on the outcome of the previous day, is known as a *transition probability*. We can arrange all such probabilities into a matrix called a *transition matrix*.

i.e.

$$\begin{array}{ll} & \text{Rain} \quad \text{Fine} \\ \begin{array}{l}\text{Rain}\\ \text{Fine}\end{array} & \begin{pmatrix} 5/8 & 3/8 \\ 3/4 & 1/4 \end{pmatrix} \end{array}$$

The entries in the first row correspond to the possible weather conditions on the following day given that it is raining at present and the entries in the second row correspond to the possible weather conditions on the following day given that it is fine at present.

Each row of a transition matrix is known as a *probability vector* and it is not difficult to see that the sum of the terms in each row must equal 1.

Worked Example 15.10

Using the data from above, find the probability distribution of the possible outcomes two days from now if it is raining today. Find the square of the transition matrix and comment on the results.

Solution 15.10

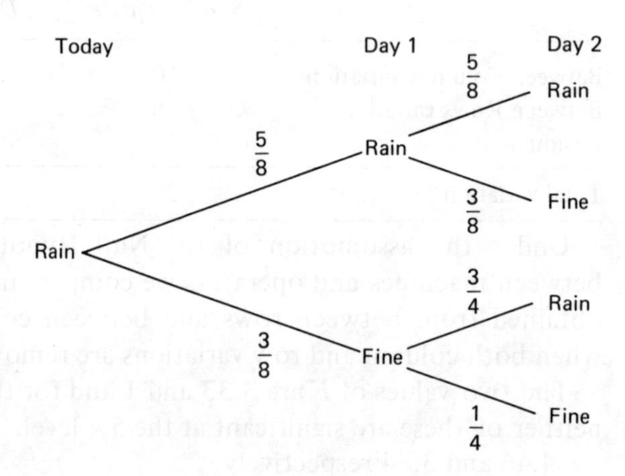

$$P(\text{Rain two days from now}) = \frac{5}{8}\cdot\frac{5}{8}+\frac{3}{8}\cdot\frac{3}{4} = \frac{43}{64}$$

$$P(\text{Fine two days from now}) = \frac{5}{8}\cdot\frac{3}{8}+\frac{3}{8}\cdot\frac{1}{4} = \frac{21}{64}$$

Outcomes	Rain	Fine
Probability	43/64	21/64

The transition matrix

$$P = \begin{pmatrix} 5/8 & 3/8 \\ 3/4 & 1/4 \end{pmatrix}$$

So $$P^2 = \begin{pmatrix} 5/8 & 3/8 \\ 3/4 & 1/4 \end{pmatrix}\begin{pmatrix} 5/8 & 3/8 \\ 3/4 & 1/4 \end{pmatrix}$$

$$= \begin{pmatrix} \frac{5}{8}\cdot\frac{5}{8}+\frac{3}{8}\cdot\frac{3}{4} & \frac{5}{8}\cdot\frac{3}{8}+\frac{3}{8}\cdot\frac{1}{4} \\ \frac{3}{4}\cdot\frac{5}{8}+\frac{1}{4}\cdot\frac{3}{4} & \frac{3}{4}\cdot\frac{3}{8}+\frac{1}{4}\cdot\frac{1}{4} \end{pmatrix}$$

$$= \begin{pmatrix} 43/64 & 21/64 \\ 21/32 & 11/32 \end{pmatrix}$$

The probabilities of 43/64 and 21/64 in the matrix are equivalent to a *two-step change* from rain to rain and from rain to fine. Similarly it can be shown that the probabilities 21/32 and 11/32 are a two-step change from fine to rain and from fine to fine.

P^2 is a *two-step transition matrix* and is usually denoted $P^{(2)}$. We can extend this idea to any positive integer n and the matrix P^n is the n-step transition matrix denoted $P^{(n)}$.

In general, if there are a finite number of outcomes or states $x_1, x_2, \ldots, x_n$ to a system then:

for the transition matrix P

p_{ij} denotes the probability that the system changes from state x_i to state x_j in one step

for the n-step transition matrix $P^{(n)}$

$p_{ij}^{(n)}$ denotes the probability that the system changes from state x_i to state x_j in n steps.

Worked Example 15.11

Mark Offchain is not a very good time-keeper. He is often late when meeting his girl friend, May Tricks. If he was on time for the previous date there is a probability of 2/5 that he will be late on the next and if he was late arriving on the previous date, there is a probability of 7/10 that he will be late on the next date. If he was late on his last date, find the probability that he will be late on the third date from now.

Solution 15.11

The transition matrix, P, is

$$P = \begin{array}{c} \\ \text{Late} \\ \text{On time} \end{array} \begin{array}{c} \begin{array}{cc} \text{Late} & \text{On time} \end{array} \\ \begin{pmatrix} 7/10 & 3/10 \\ 2/5 & 3/5 \end{pmatrix} \end{array}$$

We require $p_{11}^{(3)}$

$$P^2 = \begin{pmatrix} 7/10 & 3/10 \\ 2/5 & 3/5 \end{pmatrix} \begin{pmatrix} 7/10 & 3/10 \\ 2/5 & 3/5 \end{pmatrix} = \begin{pmatrix} 61/100 & 39/100 \\ 13/25 & 12/25 \end{pmatrix}$$

$$\text{and } P^3 = \begin{pmatrix} 7/10 & 3/10 \\ 2/5 & 3/5 \end{pmatrix} P^2 = \begin{pmatrix} 583/1000 & 417/1000 \\ 139/250 & 111/250 \end{pmatrix}$$

So $p_{11}^{(3)} = 583/1000 = 0.583$

15.11 Long Term Probabilities

In the previous example, let us suppose that before Mark decides to date May, he throws a fair coin to decide whether he will turn up late or not on the first date. The probability distribution for the possible outcomes is:

Outcome	*Late*	*On time*
Probability	1/2	1/2

and we can represent the probabilities with a probability vector

$\mathbf{p}^{(0)} = (1/2, 1/2)$

The$^{(0)}$ here represents an initial state.

The probability distribution for the second date is given by

Outcome	*Late*	*On time*
Probability	$\frac{1}{2}\cdot\frac{7}{10}+\frac{1}{2}\cdot\frac{2}{5}$	$\frac{1}{2}\cdot\frac{3}{10}+\frac{1}{2}\cdot\frac{3}{5}$
	11/20	9/20

This distribution can be represented by the probability vector $\mathbf{p}^{(1)}$ which can be determined by calculating $\mathbf{p}^{(0)}P$,

i.e. $\mathbf{p}^{(1)} = \mathbf{p}^{(0)}P = (1/2, 1/2)\begin{pmatrix} 7/10 & 3/10 \\ 2/5 & 3/5 \end{pmatrix}$

$= (11/20,\ 9/20)$

In a similar manner, the probability distribution of the outcomes or states for the third date is given by

$$\mathbf{p}^{(2)} = \mathbf{p}^{(1)}P = \mathbf{p}^{(0)}P^2$$

In general $\quad \mathbf{p}^{(n)} = \mathbf{p}^{(0)}P^n$

> The stochastic behaviour of a Markov chain is uniquely determined if the matrix P and the initial distribution are specified.

In problems involving Markov processes we are often concerned with *stable states* or *long-term probabilities*. That is as $n \to \infty$, does $\mathbf{p}^{(n)}$ approach a fixed probability vector? Before we proceed we require a definition of a *regular stochastic matrix*.

> **Definition:** A stochastic matrix P is said to be regular if for some positive integer value n, all the entries in P are greater than 0.

It is beyond the scope of this work, but it can be shown that for a regular stochastic matrix P, there exists a unique fixed probability vector p such that $\mathbf{p}P = \mathbf{p}$ and the sequence $P, P^2, P^3, \ldots$ approaches the matrix T whose rows are the component probabilities of $\mathbf{p}$.

The probability components of $\mathbf{p}$ are the long term probabilities of each state and these are independent of the initial state of the system.

Worked Example 15.12

The transition matrix P defined by Worked Example 15.11 is regular. Find the fixed probability vector $\mathbf{p}$ such that $\mathbf{p}P = \mathbf{p}$. What does P^n approach as $n \to \infty$?

Solution 15.12

Let the fixed probability vector be $(p, 1-p)$

So $(p, 1-p)\begin{pmatrix} 7/10 & 3/10 \\ 2/5 & 3/5 \end{pmatrix} = (p, 1-p)$

(must add up to 1)

$$\frac{7}{10}p + \frac{2}{5} - \frac{2}{5}p = p$$

$$\Rightarrow \frac{2}{5} = \frac{7}{10}p$$

$$p = 4/7 \Rightarrow (1-p) = 3/7$$

The fixed probability vector is therefore $(4/7, 3/7)$ and P^n approaches

$\begin{pmatrix} 4/7 & 3/7 \\ 4/7 & 3/7 \end{pmatrix}$ as $n \to \infty$

Worked Example 15.13

A Markov chain has a transition matrix defined for two states 1 and 2 as follows.

$$P = \begin{matrix} & 1 & 2 \\ 1 & 1/3 & 2/3 \\ 2 & 1/2 & 1/2 \end{matrix}$$

Initially the system is either in state 1 with a probability of 2/5 or in state 2 with a probability of 3/5.

Find (i) $p_{22}^{(2)}$, (ii) $p_1^{(2)}$, (iii) $\mathbf{p}^{(3)}$.

Solution 15.13

(i) $p_{22}^{(2)}$ is a component of the stochastic matrix $P^{(2)}$

Now $P^{(2)} = P^2 = \begin{pmatrix} 1/3 & 2/3 \\ 1/2 & 1/2 \end{pmatrix}\begin{pmatrix} 1/3 & 2/3 \\ 1/2 & 1/2 \end{pmatrix}$

$= \begin{pmatrix} 4/9 & 5/9 \\ 5/12 & 7/12 \end{pmatrix}$

So $p_{22}^{(2)} = 7/12$

(ii) $p_1^{(2)}$ is the first component of the two-step probability distribution $\mathbf{p}^{(2)}$.

Now $\mathbf{p}^{(2)} = \mathbf{p}^{(1)}P = \mathbf{p}^{(0)}P^2$

$= (2/5, 3/5)\begin{pmatrix} 4/9 & 5/9 \\ 5/12 & 7/12 \end{pmatrix}$

$= (77/180, 103/180)$

So $p_1^{(2)} = 77/180$

(iii) $\mathbf{p}^{(3)} = \mathbf{p}^{(2)}P$

$= (77/180, 103/180)\begin{pmatrix} 1/3 & 2/3 \\ 1/2 & 1/2 \end{pmatrix}$

$= (463/1080, 617/1080)$

Worked Example 15.14

Mr Peter T. Ruby, an affluent jeweller, gives his wife a ring each year. If this year he gives her an emerald ring he will give her a sapphire ring next year. If however he has given her a sapphire ring, he is just as likely to give her another sapphire ring as to give her a diamond ring, and he will not give her an emerald ring. If he gives her a diamond ring he does not repeat his choice and he is twice as likely to buy an emerald ring as he is a sapphire ring next year.

If he gives her a diamond ring this year find the probability that

(i) in two years he will give her a sapphire ring
(ii) in three years time he will not give a diamond ring.

Find also the probability that in the long term he will give her an emerald ring. You may assume that the transition matrix is regular.

Solution 15.14

The transition matrix P =

	Emerald	Sapphire	Diamond
Emerald	0	1	0
Sapphire	0	1/2	1/2
Diamond	2/3	1/3	0

(i)

$$P^2 = \begin{pmatrix} 0 & 1 & 0 \\ 0 & 1/2 & 1/2 \\ 2/3 & 1/3 & 0 \end{pmatrix}\begin{pmatrix} 0 & 1 & 0 \\ 0 & 1/2 & 1/2 \\ 2/3 & 1/3 & 0 \end{pmatrix}$$

$$= \begin{pmatrix} 0 & 1/2 & 1/2 \\ 1/3 & 5/12 & 1/4 \\ 0 & 5/6 & 1/6 \end{pmatrix}$$

The probability vector representing the initial state

$$\mathbf{p}^{(0)} = (0, 0, 1)$$

The two-step probability distribution is given by

$$\mathbf{p}^{(2)} = \mathbf{p}^{(0)}P^2$$

$$= (0, 0, 1)\begin{pmatrix} 0 & 1/2 & 1/2 \\ 1/3 & 5/12 & 1/4 \\ 0 & 5/6 & 1/6 \end{pmatrix}$$

$$= (0, 5/6, 1/6)$$

So the probability that in two years he will give his wife a sapphire ring is 5/6.

(ii) The three-step probability distribution is given by

$$\mathbf{p}^{(3)} = \mathbf{p}^{(2)}P$$

$$= (0, 5/6, 1/6)\begin{pmatrix} 0 & 1 & 0 \\ 0 & 1/2 & 1/2 \\ 2/3 & 1/3 & 0 \end{pmatrix}$$

$$= (1/9, 17/36, 5/12)$$

The probability that he does not give his wife a diamond ring is $1 - 5/12 = 7/12$.

For the last part we require the fixed probability vector $\mathbf{p}$ such that

$$\mathbf{p}P = \mathbf{p}$$

Letting the fixed probability vector be (p_1, p_2, p_3)

Then $(p_1, p_2, p_3)\begin{pmatrix} 0 & 1 & 0 \\ 0 & 1/2 & 1/2 \\ 2/3 & 1/3 & 0 \end{pmatrix} = (p_1, p_2, p_3)$

So $\left(\frac{2}{3}p_3, p_1 + \frac{1}{2}p_2 + \frac{1}{3}p_3, \frac{1}{2}p_2\right) = (p_1, p_2, p_3)$

Equating terms $p_3 = 3p_1/2$ and $p_2 = 2p_3, p_2 = 3p_1$

and using the fact that $p_1 + p_2 + p_3 = 1$

$$p_1 + 3p_1 + \frac{3p_1}{2} = 1$$

$$\Rightarrow p_1 = 2/11$$

So the long term probability of giving his wife an emerald ring is 2/11.

Worked Example 15.15

A Markov chain on states $\{1, 2, 3, 4, 5\}$ has transition matrix

$$P = \begin{pmatrix} 2/3 & 2/9 & 2/27 & 2/81 & 1/81 \\ 2/3 & 2/9 & 2/27 & 2/81 & 1/81 \\ 0 & 2/3 & 2/9 & 2/27 & 1/27 \\ 0 & 0 & 2/3 & 2/9 & 1/9 \\ 0 & 0 & 0 & 2/3 & 1/3 \end{pmatrix}$$

(i) Find $p_{22}^{(2)}$.

(ii) For what values of n are $p_{ij}^{(n)} > 0$ for all i, j? Justify your argument carefully.

(iii) Given that $p_{11}^{(n)}$ converges to 16/31 as $n \to \infty$, find the long-term probability of being in state 2. (OLE)

Solution 15.15

(i) $p_{22}^{(2)}$, a term in P^2, can be determined by multiplying together row 2 and column 2 of P.

$$p_{22}^{(2)} = \frac{2}{3}\cdot\frac{2}{9} + \frac{2}{9}\cdot\frac{2}{9} + \frac{2}{27}\cdot\frac{2}{3}$$

$$= 20/81$$

(ii) In calculating P^2, three of the six zeros in the transition matrix can be removed:

$$p_{31}^{(2)} = p_{32}^{(1)} \cdot p_{21}^{(1)} > 0$$

$$p_{42}^{(2)} = p_{43}^{(1)} \cdot p_{32}^{(1)} > 0$$

and $$p_{53}^{(2)} = p_{54}^{(1)} \cdot p_{43}^{(1)} > 0$$

But $p_{41}^{(2)} = p_{51}^{(2)} = p_{52}^{(2)} = 0$

So

$$P^2 = \begin{pmatrix} & & \\ & p_{ij}^{(2)} > 0 & \\ 0 & & \\ 0 \quad 0 & & \end{pmatrix}$$

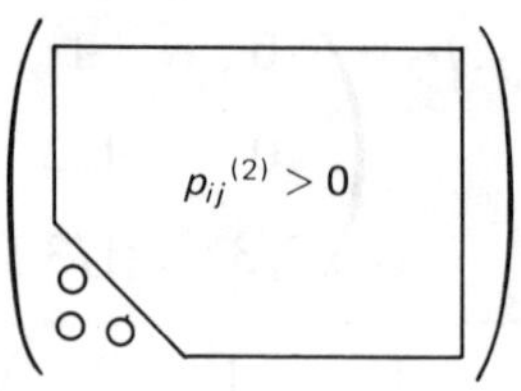

In calculating P^3, a further two zeros are removed

$$p_{41}^{(3)} = p_{43}^{(1)} . p_{31}^{(2)} > 0$$

and
$$p_{52}^{(3)} = p_{54}^{(1)} . p_{42}^{(2)} > 0$$

But $p_{51}^{(3)} = p_{54}^{(1)} . p_{41}^{(2)} + p_{55}^{(1)} . p_{51}^{(2)} = 0$

So

$$P^3 = \begin{pmatrix} & & \\ & p_{ij}^{(3)} > 0 & \\ & & \\ 0 & & \end{pmatrix}$$

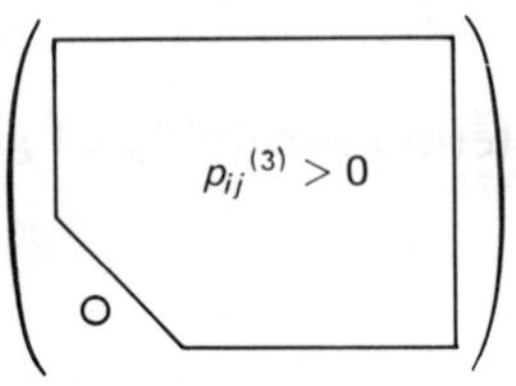

The final zero is 'eliminated' in P^4 as

$$p_{51}^{(4)} = p_{54}^{(1)} . p_{41}^{(3)} > 0$$

The values of n must therefore exceed 3.

(iii) $\lim_{n \to \infty} p_{11}^{(n)} = \dfrac{16}{31}$

This limit is independent of the initial state $\Rightarrow p_1^{(n)} = 16/31$.

This is the first component of the fixed probability vector. We now require to find $p_2^{(n)}$.

Letting $p_2^{(n)} = x$

$$(16/31, x, p_3^{(n)}, p_4^{(n)}, p_5^{(n)})P = (16/31, x, p_3^{(n)}, p_4^{(n)}, p_5^{(n)})$$

$$\Rightarrow \frac{16}{31} \cdot \frac{2}{3} + x \cdot \frac{2}{3} = \frac{16}{31}$$

$$\Rightarrow \frac{16}{31} + x = \frac{24}{31}$$

$$\Rightarrow x = 8/31$$

So the long term probability of being in state 2 is 8/31.

15.12 Exercises

Exercise 15.1

A large food processing firm is considering introducing a new recipe for its ice cream. In a preliminary trial, a panel of 11 tasters were asked to score ice cream made from both the existing and the new recipe for sweetness. The results, on a scale from 0 to 100 with the sweeter ice cream being given the higher score, were as follows:

Taster	*A*	*B*	*C*	*D*	*E*	*F*	*G*	*H*	*I*	*J*	*K*
Existing recipe	88	35	67	17	24	32	8	44	73	47	25
New recipe	94	49	66	82	25	96	14	56	27	44	79

Use the sign test, at the 5% significance level, to test whether the new recipe is sweeter than the existing one. (AEB 1984)

Exercise 15.2

A chemist determined the pH values of the surface soil and subsoil at eleven different locations in Devon, with the results in Table 15.14. Perform the Wilcoxon test to determine whether these results support the hypothesis that subsoil pH tends to be higher than surface soil pH.

Table 15.14 pH values of soil types

Location	*Surface pH*	*Subsoil pH*
1	6.57	8.34
2	6.77	6.13
3	6.53	6.32
4	6.71	8.30
5	6.72	8.44
6	6.01	6.80
7	4.99	4.42
8	5.49	7.90
9	5.56	5.20
10	5.32	5.32
11	5.92	5.13

(L)

Exercise 15.3

Ten boys and ten girls were chosen at random from a group of children of comparable age in a school. The weekly pocket money of each child was recorded. The investigator found the median (£1.95) of the complete set of data and tabulated the data as in Table 15.15. Using a non-parametric test, test whether the boys' pocket money differs significantly from the girls'. (OLE)

Table 15.15

		Weekly pocket money (£)	*Number in group*
Boys.	Less than median	0.4, 0.9, 1.1, 1.4, 1.5, 1.6, 1.7	7
	Greater than median	2.3, 3.5, 8.6	3
Girls.	Less than median	1.0, 1.0, 1.8	3
	Greater than median	2.1, 2.4, 3.5, 5.5, 6.0, 6.3, 7.3	7

Exercise 15.4

(a) For the case of two paired samples, explain briefly

(i) the circumstances under which the t-test would be appropriate.
(ii) the relative advantages of the sign test and of the Wilcoxon matched-pairs signed rank test.

(b) An investigator is interested in the effects of alcohol on the accuracy of throwing of a dart. Eight volunteers each throw a single dart, aiming at the centre of the dart-board. They then each have a pint of beer and then throw the dart once more. The distances (in cm) from the centre of the dart-board are recorded in Table 15.16.

Table 15.16

Volunteer	*1*	*2*	*3*	*4*	*5*	*6*	*7*	*8*
Before beer	5.0	2.4	6.9	0.3	4.1	3.2	2.0	4.2
After beer	3.7	7.6	7.1	5.3	4.7	1.1	2.4	4.9

(i) State explicity suitable null and alternative hypotheses.
(ii) Using the sign test, determine the set of values of α for which your null hypothesis would be rejected at the α% significance level.
(iii) Using the Wilcoxon matched-pairs signed rank test, carry out a test of the null hypothesis at the 5% significance level, and state your conclusions. (UCLES)

Exercise 15.5

An experiment was conducted to study the effects of various diets on pigs. A total of 24 similar pigs were selected and randomly allocated to one of the five groups such that the control group, which was fed a normal diet, had 8 pigs and each of the other groups, to which the new diets were given, had 4 pigs each. After a fixed time the gains in mass, in kilograms, of the pigs were measured. Unfortunately by this time two pigs had died, one which was on diet A and one which was on diet C. The gains in mass of the remaining pigs are recorded in Table 15.17.

Table 15.17

Diets	*Gains in mass (kg)*
Normal	23.1, 9.8, 15.5, 22.6, 14.6, 11.2, 15.7, 10.5
A	21.9, 13.2, 19.7
B	16.5, 22.8, 18.3, 31.0
C	30.9, 21.9, 29.8
D	21.0, 25.4, 21.5, 21.2

Use a one-factor analysis of variance to test, at the 5% level of significance, for differences between diets.

What further information would you require about the dead pigs and how might this affect the conclusions of your analysis? (AEB 1982)

Exercise 15.6

Information about the current state of a complex industrial process is displayed on a control panel which is monitored by a technician. In order to find the best display for the instruments on the control panel three different arrangements were tested by simulating an emergency and observing the reaction times of five different technicians. The results, in seconds, are given in Table 15.18.

Table 15.18

		Technician				
		P	*Q*	*R*	*S*	*T*
	A	2.4	3.3	1.9	3.6	2.7
Arrangement	B	3.7	3.2	2.7	3.9	4.4
	C	4.2	4.6	3.9	3.8	4.5

Carry out an analysis of variance and test for differences between technicians and between arrangements at the 5% significance level. Currently arrangement C is used and it is suggested that this be replaced by arrangement A. Comment, briefly, on this suggestion and on what further information you would find useful before coming to a definite decision. (AEB 1984)

Exercise 15.7

(a) As part of a project to improve the steerability of trucks a manufacturer took three trucks of the same model and fitted them with soft, standard and hard front springs respectively. The turning radius (the radius of the circle in which the truck could turn full circle) was measured for each truck using a variety of drivers, speeds and surface conditions. Use the information in Table 15.19 to test for a difference between springs at the 5% significance level.

Table 15.19

Source	*Sum of squares*	*Degrees of freedom*
Between springs	37.9	2
Within springs	75.6	18
Total	113.5	20

(b) A statistician suggested that the experiment would be improved if the same truck was used all the time with the front springs being changed as necessary and if the speed of the truck was controlled.
The results in Table 15.20 for turning circle (metres) were obtained.

Table 15.20

		Springs		
		Soft	*Standard*	*Hard*
Speed	15 km/h	42	43	39
	25 km/h	48	50	48

Carry out a two factor analysis of variance and test at the 5% significance level for differences between springs and between speeds. You may assume that the total sum of squares about the mean (SS_T) is 92.

(c) Compare the two experiments and suggest a further improvement to the design. (AEB 1989)

Exercise 15.8

Peter Vector, on holiday in the city of Stoke Astic, stayed at the Hotel Markoff. On one particular day he set off from his hotel to see the four 'sights' of the city,

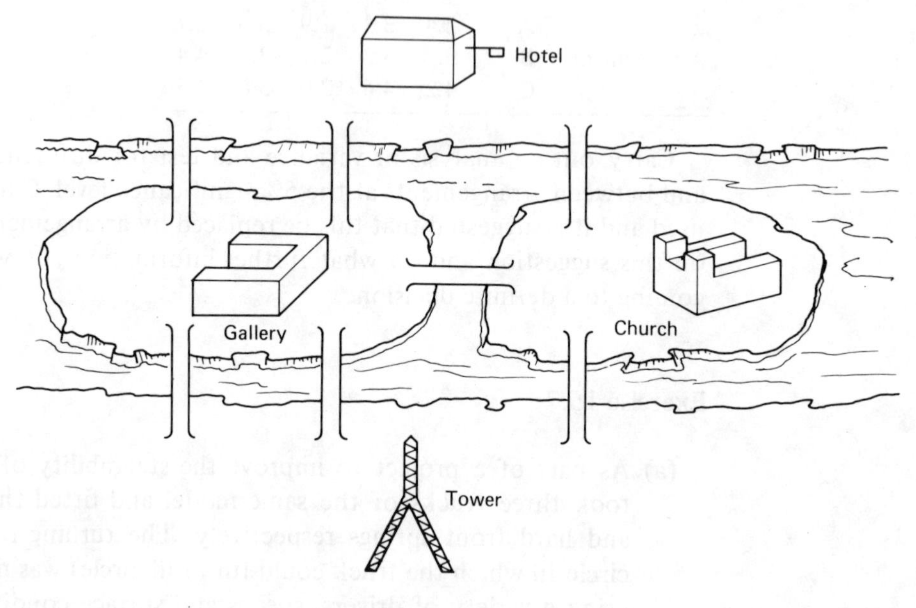

Figure 15.1

the Hotel Markoff, the Gallery, the Church and the Tower, on foot. In order to get from one sight to another he crosses one of the seven bridges, but unfortunately he has little sense of direction and every bridge from a given sight is equally likely to be chosen. Given that he spends one hour at every sight he arrives at irrespective of whether he has already visited it, and that he leaves the Hotel Markoff at 9.00 am to cross one of the bridges:

(i) Write down the transition matrix.
(ii) Find the probability that he is at the Church at 11.30 am.
(iii) Given that the long-term probability of being at the hotel in a given hour is 3/14, find the long term probability of being at the Gallery.

(You may assume that travelling time is negligible.)

Exercise 15.9

A Markov chain on the states $\{1, 2, 3, 4, 5\}$ has transition matrix (p_{ij}), where, for $1 \leqslant i \leqslant 3$, $p_{ii} = x$, $p_{i,i+1} = y$, $p_{i5} = 1 - x - y$ and $p_{44} = x$, $p_{45} = 1 - x$, $p_{51} = 1$, all other entries being zero.

(i) Find the distribution of the state after two steps, starting in state 1.
(ii) Deduce that $p_{12}^{(3)} = 3x^2y + y - xy - y^2$.

(iii) Given that (8/19, 4/19, 2/19, 1/19, 4/19) are the long term probabilities of being in the various states, find the values of x and y, and the mean time to return to state 5 if we are in state 5 now. (OLE)

Exercise 15.10

Secret tunnels connect the four rooms on the ground floor of a castle. There are 4 tunnels between rooms A and B, 3 tunnels between A and C, 2 tunnels between B and D, and 3 tunnels between C and D. Robert and Daniel are initially in rooms A and B respectively; once every minute, John blows a whistle, whereupon Robert and Daniel each select one tunnel at random in the room they are in, and travel along it to its end in another room. Thus each of their positions at times 0, 1, 2, . . . form a four-state Markov chain; write down its transition matrix.

Find the probability distributions of the rooms Robert and Daniel are in after two movements.

Explain why they will never be in the same room at the same time.

Show that, if Peter enters one of the four rooms at random, and moves according to the same rules, the long term probability of his being in a given room is proportional to the number of tunnels in that room. (OLE)

16 Tables

16.1 Areas in the Tail of the Normal Distribution

The function tabulates is $1 - F(z) = P(Z > z)$

where $z = \dfrac{x - \mu}{\sigma}$

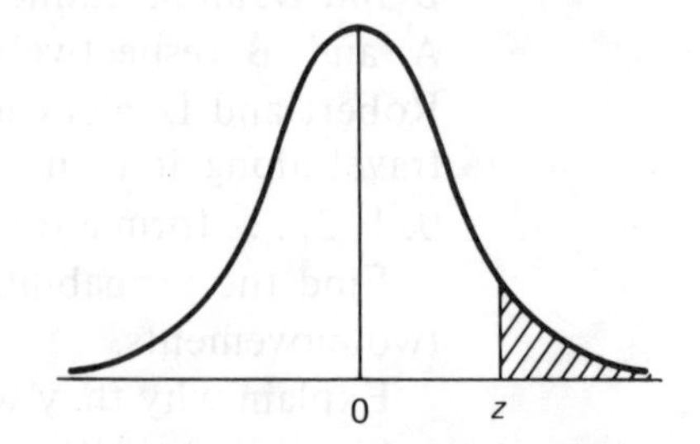

z	0.00	0.01	0.02	0.03	0.04	0.05	0.06	0.07	0.08	0.09
0.0	0.5000	0.4960	0.4920	0.4880	0.4840	0.4801	0.4761	0.4721	0.4681	0.4641
0.1	0.4602	0.4562	0.4522	0.4483	0.4443	0.4404	0.4364	0.4325	0.4286	0.4247
0.2	0.4207	0.4168	0.4129	0.4090	0.4052	0.4013	0.3974	0.3936	0.3897	0.3859
0.3	0.3821	0.3783	0.3745	0.3707	0.3669	0.3632	0.3594	0.3557	0.3520	0.3483
0.4	0.3446	0.3409	0.3372	0.3336	0.3300	0.3264	0.3228	0.3192	0.3156	0.3121
0.5	0.3085	0.3050	0.3015	0.2981	0.2946	0.2912	0.2877	0.2843	0.2810	0.2776
0.6	0.2743	0.2709	0.2676	0.2643	0.2611	0.2578	0.2546	0.2514	0.2483	0.2451
0.7	0.2420	0.2389	0.2358	0.2327	0.2296	0.2266	0.2236	0.2206	0.2177	0.2148
0.8	0.2119	0.2090	0.2061	0.2033	0.2005	0.1977	0.1949	0.1922	0.1894	0.1867
0.9	0.1841	0.1814	0.1788	0.1762	0.1736	0.1711	0.1685	0.1660	0.1635	0.1611
1.0	0.1587	0.1562	0.1539	0.1515	0.1492	0.1469	0.1446	0.1423	0.1401	0.1379
1.1	0.1357	0.1335	0.1314	0.1292	0.1271	0.1251	0.1230	0.1210	0.1190	0.1170
1.2	0.1151	0.1131	0.1112	0.1093	0.1075	0.1056	0.1038	0.1020	0.1003	0.0985
1.3	0.0968	0.0951	0.0934	0.0918	0.0901	0.0885	0.0869	0.0853	0.0838	0.0823
1.4	0.0808	0.0793	0.0778	0.0764	0.0749	0.0735	0.0721	0.0708	0.0694	0.0681
1.5	0.0668	0.0655	0.0643	0.0630	0.0618	0.0606	0.0594	0.0582	0.0571	0.0559
1.6	0.0548	0.0537	0.0526	0.0516	0.0505	0.0495	0.0485	0.0475	0.0465	0.0455
1.7	0.0446	0.0436	0.0427	0.0418	0.0409	0.0401	0.0392	0.0384	0.0375	0.0367
1.8	0.0359	0.0351	0.0344	0.0336	0.0329	0.0322	0.0314	0.0307	0.0301	0.0294
1.9	0.0287	0.0281	0.0274	0.0268	0.0262	0.0256	0.0250	0.0244	0.0239	0.0233
2.0	0.02275	0.02222	0.02169	0.02118	0.02068	0.02018	0.01970	0.01923	0.01876	0.01831
2.1	0.01786	0.01743	0.01700	0.01659	0.01618	0.01578	0.01539	0.01500	0.01463	0.01426
2.2	0.01390	0.01355	0.01321	0.01287	0.01255	0.01222	0.01191	0.01160	0.01130	0.01101
2.3	0.01072	0.01044	0.01017	0.00990	0.00964	0.00939	0.00914	0.00889	0.00866	0.00842
2.4	0.00820	0.00798	0.00776	0.00755	0.00734	0.00714	0.00695	0.00676	0.00657	0.00639
2.5	0.00621	0.00604	0.00587	0.00570	0.00554	0.00539	0.00523	0.00508	0.00494	0.00480
2.6	0.00466	0.00453	0.00440	0.00427	0.00415	0.00402	0.00391	0.00379	0.00368	0.00357
2.7	0.00347	0.00336	0.00326	0.00317	0.00307	0.00298	0.00289	0.00280	0.00272	0.00264
2.8	0.00256	0.00248	0.00240	0.00233	0.00226	0.00219	0.00212	0.00205	0.00199	0.00193
2.9	0.00187	0.00181	0.00175	0.00169	0.00164	0.00159	0.00154	0.00149	0.00144	0.00139
3.0	0.00135									
3.1	0.00097									
3.2	0.00069									
3.3	0.00048									
3.4	0.00034									
3.5	0.00023									
3.6	0.00016									
3.7	0.00011									
3.8	0.00007									
3.9	0.00005									
4.0	0.00003									

16.2 Percentage Points of the t-Distribution

The table gives the value of t_ν for different percentage points of the t distribution for ν degrees of freedom.

Note. The tabulation is for one tail only, i.e. for positive values of t. For $|t|$ the column headings for α must be doubled.

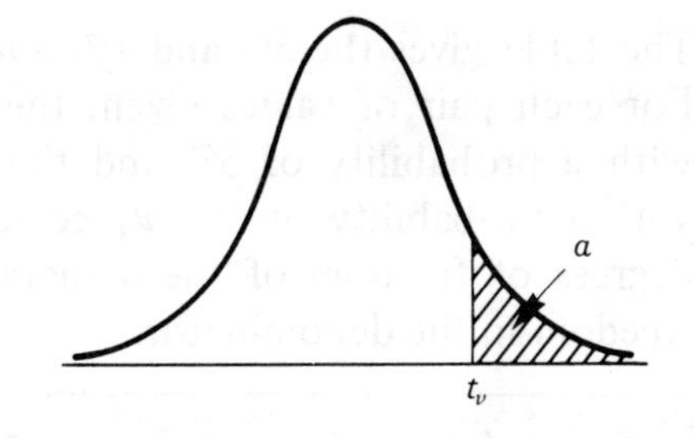

Percentage points (α)	*0.05*	*0.025*	*0.01*	*0.005*
ν = 1	6.314	12.706	31.821	63.657
2	2.920	4.303	6.965	9.925
3	2.353	3.182	4.541	5.841
4	2.132	2.776	3.747	4.604
5	2.015	2.571	3.365	4.032
6	1.943	2.447	3.143	3.707
7	1.895	2.365	2.998	3.499
8	1.860	2.306	2.896	3.355
9	1.833	2.262	2.821	3.250
10	1.812	2.228	2.764	3.169
11	1.796	2.201	2.718	3.106
12	1.782	2.179	2.681	3.055
13	1.771	2.160	2.650	3.012
14	1.761	2.145	2.624	2.977
15	1.753	2.131	2.602	2.947
16	1.746	2.120	2.583	2.921
17	1.740	2.110	2.567	2.898
18	1.734	2.101	2.552	2.878
19	1.729	2.093	2.539	2.861
20	1.725	2.086	2.528	2.845
21	1.721	2.080	2.518	2.831
22	1.717	2.074	2.508	2.819
23	1.714	2.069	2.500	2.807
24	1.711	2.064	2.492	2.797
25	1.708	2.060	2.485	2.787
26	1.706	2.056	2.479	2.779
27	1.703	2.052	2.473	2.771
28	1.701	2.048	2.467	2.763
29	1.699	2.045	2.462	2.756
30	1.697	2.042	2.457	2.750
40	1.684	2.021	2.423	2.704
60	1.671	2.000	2.390	2.660
120	1.658	1.980	2.358	2.617
∞	1.645	1.960	2.326	2.576

This table is taken from Table III of Fisher & Yates: *Statistical Tables for Biological, Agricultural and Medical Research*, published by Oliver & Boyd Ltd., Edinburgh, and by permission of the authors and publishers and also from Table 12 of Biometrika *Tables for Statisticians*, Volume 1, by permission of the Biometrika Trustees.

16.3 Percentage Points of the F-Distribution

The table gives the 5% and 1% points of the F-distribution. For each pair of values given, the upper figure is exceeded with a probability of 5% and the lower figure is exceeded with a probability of 1%. ν_1 corresponds to the number of degrees of freedom of the numerator and ν_2 the degrees of freedom in the denominator.

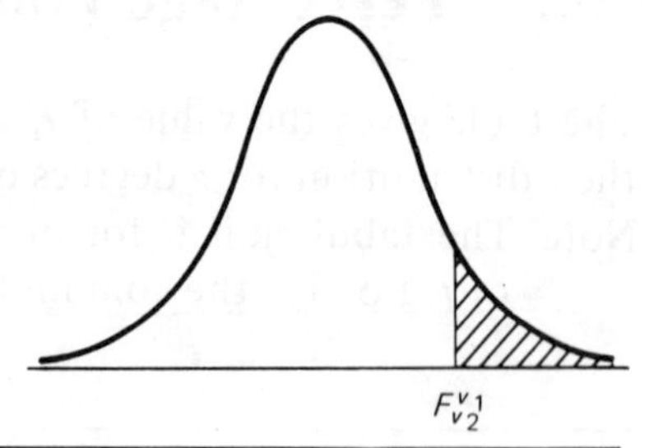

$\nu_2 \backslash \nu_1$	1	2	3	4	5	6	7	8	10	12	24	∞
1	161.4	199.5	215.7	224.6	230.2	234.0	236.8	238.9	241.9	243.9	249.0	254.3
	4052	5000	5403	5625	5764	5859	5928	5981	6056	6106	6235	6366
2	18.5	19.0	19.2	19.2	19.3	19.3	19.4	19.4	19.4	19.4	19.5	19.5
	98.5	99.0	99.2	99.2	99.3	99.3	99.4	99.4	99.4	99.4	99.5	99.5
3	10.13	9.55	9.28	9.12	9.01	8.94	8.89	8.85	8.79	8.74	8.64	8.53
	34.1	30.8	29.5	28.7	28.2	27.9	27.7	27.5	27.2	27.1	26.6	26.1
4	7.71	6.94	6.59	6.39	6.26	6.16	6.09	6.04	5.96	5.91	5.77	5.63
	21.2	18.0	16.7	16.0	15.5	15.2	15.0	14.8	14.5	14.4	13.9	13.5
5	6.61	5.79	5.41	5.19	5.05	4.95	4.88	4.82	4.74	4.68	4.53	4.36
	16.26	13.27	12.06	11.39	10.97	10.67	10.46	10.29	10.05	9.89	9.47	9.02
6	5.99	5.14	4.76	4.53	4.39	4.28	4.21	4.15	4.06	4.00	3.84	3.67
	13.74	10.92	9.78	9.15	8.75	8.47	8.26	8.10	7.87	7.72	7.31	6.88
7	5.59	4.74	4.35	4.12	3.97	3.87	3.79	3.73	3.64	3.57	3.41	3.23
	12.25	9.55	8.45	7.85	7.46	7.19	6.99	6.84	6.62	6.47	6.07	5.65
8	5.32	4.46	4.07	3.84	3.69	3.58	3.50	3.44	3.35	3.28	3.12	2.93
	11.26	8.65	7.59	7.01	6.63	6.37	6.18	6.03	5.81	5.67	5.28	4.86
9	5.12	4.26	3.86	3.63	3.48	3.37	3.29	3.23	3.14	3.07	2.90	2.71
	10.56	8.02	6.99	6.42	6.06	5.80	5.61	5.47	5.26	5.11	4.73	4.31
10	4.96	4.10	3.71	3.48	3.33	3.22	3.14	3.07	2.98	2.91	2.74	2.54
	10.04	7.56	6.55	5.99	5.64	5.39	5.20	5.06	4.85	4.71	4.33	3.91
11	4.84	3.98	3.59	3.36	3.20	3.09	3.01	2.95	2.85	2.79	2.61	2.40
	9.65	7.21	6.22	5.67	5.32	5.07	4.89	4.74	4.54	4.40	4.02	3.60
12	4.75	3.89	3.49	3.26	3.11	3.00	2.91	2.85	2.75	2.69	2.51	2.30
	9.33	6.93	5.95	5.41	5.06	4.82	4.64	4.50	4.30	4.16	3.78	3.36
13	4.67	3.81	3.41	3.18	3.03	2.92	2.83	2.77	2.67	2.60	2.42	2.21
	9.07	6.70	5.74	5.21	4.86	4.62	4.44	4.30	4.10	3.96	3.59	3.17
14	4.60	3.74	3.34	3.11	2.96	2.85	2.76	2.70	2.60	2.53	2.35	2.13
	8.86	6.51	5.56	5.04	4.70	4.46	4.28	4.14	3.94	3.80	3.43	3.00
16	4.49	3.63	3.24	3.01	2.85	2.74	2.66	2.59	2.49	2.42	2.24	2.01
	8.53	6.23	5.29	4.77	4.44	4.20	4.03	3.89	3.69	3.55	3.18	2.75
18	4.41	3.55	3.16	2.93	2.77	2.66	2.58	2.51	2.41	2.34	2.15	1.92
	8.29	6.01	5.09	4.58	4.25	4.01	3.84	3.71	3.51	3.37	3.00	2.57
20	4.35	3.49	3.10	2.87	2.71	2.60	2.51	2.45	2.35	2.28	2.08	1.84
	8.10	5.85	4.94	4.43	4.10	3.87	3.70	3.56	3.37	3.23	2.86	2.42
22	4.30	3.44	3.05	2.82	2.66	2.55	2.46	2.40	2.30	2.23	2.03	1.78
	7.95	5.72	4.82	4.31	3.99	3.76	3.59	3.45	3.26	3.12	2.75	2.31
24	4.26	3.40	3.01	2.78	2.62	2.51	2.42	2.36	2.25	2.18	1.98	1.73
	7.82	5.61	4.72	4.22	3.90	3.67	3.50	3.36	3.17	3.03	2.66	2.21
∞	3.84	3.00	2.60	2.37	2.21	2.10	2.01	1.94	1.83	1.75	1.52	1.00
	6.63	4.61	3.78	3.32	3.02	2.80	2.64	2.51	2.32	2.18	1.79	1.00

This table is taken from Table V of Fisher & Yates: *Statistical Tables for Biological, Agricultural and Medical Research*, published by Oliver & Boyd Ltd., Edinburgh, and by permission of the authors and publishers and also from Table 18 of *Biometrika Tables for Statisticians*, Volume 1, by permission of the Biometrika Trustees.

16.4 Percentage Points of the χ^2-Distribution

Table 16.4 Percentage points of the χ^2-distribution which are exceeded with a probability α for ν degrees of freedom.

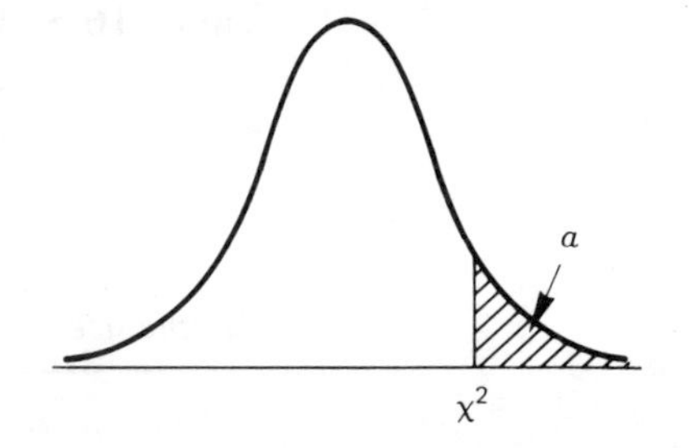

$\alpha =$	0.975	0.95	0.05	0.025	0.01	0.005	0.001	$= \alpha$
$\nu =$ 1	0.0^3982	0.00393	3.841	5.024	6.635	7.879	10.827	$\nu =$ 1
2	0.0506	0.103	5.991	7.378	9.210	10.597	13.815	2
3	0.216	0.352	7.815	9.348	11.345	12.838	16.268	3
4	0.484	0.711	9.488	11.143	13.277	14.860	18.465	4
5	0.831	1.145	11.070	12.832	15.086	16.750	20.517	5
6	1.237	1.635	12.592	14.449	16.812	18.548	22.457	6
7	1.690	2.167	14.067	16.013	18.475	20.278	24.322	7
8	2.180	2.733	15.507	17.535	20.090	21.955	26.125	8
9	2.700	3.325	16.919	19.023	21.666	23.589	27.877	9
10	3.247	3.940	18.307	20.483	23.209	25.188	29.588	10
11	3.816	4.575	19.675	21.920	24.725	26.757	31.264	11
12	4.404	5.226	21.026	23.337	26.217	28.300	32.909	12
13	5.009	5.892	22.362	24.736	27.688	29.819	34.528	13
14	5.629	6.571	23.685	26.119	29.141	31.319	36.123	14
15	6.262	7.261	24.996	27.488	30.578	32.801	37.697	15
16	6.908	7.962	26.296	28.845	32.000	34.267	39.252	16
17	7.564	8.672	27.587	30.191	33.409	35.718	40.790	17
18	8.231	9.390	28.869	31.526	34.805	37.156	42.312	18
19	8.907	10.117	30.144	32.852	36.191	38.582	43.820	19
20	9.591	10.851	31.410	34.170	37.566	39.997	45.315	20
21	10.283	11.591	32.671	35.479	38.932	41.401	46.797	21
22	10.982	12.338	33.924	36.781	40.289	42.796	48.268	22
23	11.688	13.091	35.172	38.076	41.638	44.181	49.728	23
24	12.401	13.848	36.415	39.364	42.980	45.558	51.179	24
25	13.120	14.611	37.652	40.646	44.314	46.928	52.620	25
26	13.844	15.379	38.885	41.923	45.642	48.290	54.052	26
27	14.573	16.151	40.113	43.194	46.963	49.645	55.476	27
28	15.308	16.928	41.337	44.461	48.278	50.993	56.893	28
29	16.047	17.708	42.557	45.722	49.588	52.336	58.302	29
30	16.791	18.493	43.773	46.979	50.892	53.672	59.703	30
40	24.433	26.509	55.759	59.342	63.691	66.766	73.402	40
50	32.357	34.764	67.505	71.420	76.154	79.490	86.661	50
60	40.482	43.188	79.082	83.298	88.379	91.952	99.607	60
70	48.758	51.739	90.531	95.023	100.425	104.215	112.317	70
80	57.153	60.391	101.880	106.629	112.329	116.321	124.839	80
90	65.646	69.126	113.145	118.136	124.116	128.299	137.208	90
100	74.222	77.929	124.342	129.561	135.807	140.170	149.449	100

This table is taken by consent from *Statistical Tables for Biological, Agricultural, and Medical Research*, by R. A. Fisher and F. Yates, published by Oliver and Boyd, Edinburgh, and from Table 8 of *Biometrika Tables for Statisticians*, Vol. 1, by permission of the Biometrika Trustees.

16.5 **Critical Values for the Wilcoxon Signed Rank Test**

Table 16.5 Critical values for non-parametric tests. Critical values of T for the Wilcoxon signed rank test

	5% level of significance	
Sample size	*One tail*	*Two tail*
6	2	1
7	4	2
8	6	4
9	8	6
10	11	8
11	14	11
12	17	14
13	21	17
14	26	21
15	30	25

For a given sample size, if the calculated value of T is less than or equal to the table value for the appropriate test, then the result is significant.

16.6 **Critical Values for the Wilcoxon Rank Sum Test**

Table 16.6 Critical values of R for the Wilcoxon rank sum test

$n \backslash m$	*4*	*5*	*6*	*7*	*8*	*9*	*10*
4	10 26 11 25	11 29 12 28	12 32 13 31	13 35 14 34	14 38 15 37	15 41 16 40	15 45 17 43
5		17 38 19 36	18 42 20 40	20 45 21 44	21 49 23 47	22 53 24 51	23 57 26 54
6			26 52 28 50	27 57 29 55	29 61 31 59	31 65 33 63	32 70 35 67
7				36 69 39 66	38 74 41 71	40 79 43 76	42 84 45 81
8					49 87 51 85	51 93 54 90	53 99 56 96
9						63 108 66 105	65 115 69 111
10							78 132 82 128

For each value of n and m there are four numbers. The top pair are the critical values for a 5% level of significance, two-tailed test. The lower pair are the critical values for a 5% level of significance one-tailed test.

16.7 **Critical Values for the Mann–Whitney Test**

Table 16.7 Critical values of U for the Mann–Whitney test

$n \backslash m$	*4*	*5*	*6*	*7*	*8*	*9*	*10*
4	0 16 1 15	1 19 2 18	2 22 3 21	3 25 4 24	4 28 5 27	5 31 6 30	5 35 7 33
5		2 23 4 21	3 27 5 25	5 30 6 29	6 34 8 32	7 38 9 36	8 42 11 39
6			5 31 7 29	6 36 8 34	8 40 10 38	10 44 12 42	11 49 14 46
7				8 41 11 38	10 46 13 43	12 51 15 48	14 56 17 53
8					13 51 15 49	15 57 18 54	17 63 20 60
9						18 63 21 60	20 70 24 66
10							23 77 27 73

For each value of n and m there are four numbers. The top pair are the critical values for a 5% level of significance, two-tailed test. The lower pair are the critical values for a 5% level of significance one-tailed test.

Answers and Hints on Solutions to Exercises

Chapter 1

1. (i) continuous; (ii) discrete;
   (iii) discrete; (iv) continuous.
5. (b) 0.07
8. 10%
9. 2.45–2.95.
10. 23.12

Chapter 2

1. (i) 56.5; (ii) 57.5, 18.3.
2. (i) 27 yr; (ii) 13%.
3. (i) 3.3 kg; (ii) 0.7 kg; (iii) 68%.
4. 2°C, 0.47°C.
5. 35, 16.8, 60, 67.2.
6. (a) (i) £124; (ii) £38.07; (b) £121.53.
7. 4, 1, 0, 5, 1, 10, 9, 15, 3, 4, 2. (i) 13.4 hours; (ii) 2.7 hours.
8. (a) 17, 6. (b) (i) 96, 24; (ii) 10, 4. (c) (i) John = 61, Peter = 62; Peter performs better; (ii) 78.
9. (a) (i) m + k, s, (ii) mk, ms, 11, 4, {35, 45, 55, 75, 90}; (b) 5.6.
10. 6.5, 6.6, 5.7.
11. (a) 0.78; (b) 4.44.
12. (ii) 4.4, 0.906; (iii) 3.45–3.95; (iv) 4.53.
13. 92.55, 13.29.
14. (b) 5 ft $7\frac{1}{2}$ in, 2 in.
15. (i) 31.2; (ii) (a) 18.4; (ii) (b) 23.2.
16. (a) (i) 105, (ii) £40, £50; (b) approximately 112.
17. (i) 242; (ii) 277.

Chapter 3

1. 5/18.
2. 16/25. Find two ways of expressing $P(A \cap B)$.
3. 15.
4. 17. Form an inequality involving $(5/6)^n$.
5. 0.169. Bayes' theorem.
6. 1/2.

7. $3\pi/32$.
8. 5040, 0.75. Find the number of arrangements where the D's are together.
9. 252, 5/32. Look for a Pascal triangle pattern for the number of ways of reaching a junction.
10. (i) 1/11; (ii) 21/55; (iii) 3/11; (iv) 2/55; (v) 36/55; (vi) 5/11.
11. (b) (i) 0.375. (ii) Neither mutually exclusive nor independent.
12. 0.273. Bayes' theorem.
13. (i) 0.3846; (ii) 0.1716; (iii) 0.1506.
14. $y = (3x - 1)/2$. A diagram would be useful. $1/3 \leqslant x \leqslant 1/2$; 1/3.
15. (a) (i) 495, (ii) 28/55. Combinations. (b) (i) 0.82, (ii) 0.341. $P(X \cap Y) = P(X).P(Y)$
16. 60. (a) 1/60; (b) 2/5; (c) 1/25.
17. (i) 1/6; (ii) 1/9; (iii) $(2/3)^{r-1}$ (1/6); 1/2. Sum of an infinite geometric progression.
18. (a) 12!; (b) 1/6; (c) 3/22 $\left[\frac{1}{6} \times \left(\frac{11! - 2 \times 10!}{11!}\right)\right]$. (a) 11!; (b) 2/11;

    (c) 8/55 $\left[\frac{2}{11} \times \left(\frac{10! - 2 \times 9!}{10!}\right)\right]$.

Chapter 4

1. (i) 13/36; (ii) 12.25; (iii) 79.97; (iv) 0.615.
2. 2.55, 1.45.
3. (i) £0.55; (ii) £1.10.
4. 3.375, 4.23.
5. (i) 4.65; 0.927; (ii) 5.95; 8.347.

9.

$Y \backslash X$	1	2	3	4	5	6	$P(Y)$
1	1/6	1/36	1/36	1/36	1/36	1/36	11/36
2	0	5/36	1/36	1/36	1/36	1/36	9/36
3	0	0	4/36	1/36	1/36	1/36	7/36
4	0	0	0	3/36	1/36	1/36	5/36
5	0	0	0	0	2/36	1/36	3/36
6	0	0	0	0	0	1/36	1/36
$P(X)$	1/6	1/6	1/6	1/6	1/6	1/6	

(i) 2.53; (ii) 1.97; (iii) 10.31; (iv) 1.46.

10. (i) 2, 0.2; (ii) 0.944, 0.028, 2, 0.056.
11. 0.965, 7/8. Construct the joint distribution table – there should be 7 non-zero entries.
12. 5/3, 5/9, 7/3, 5/9. (ii) Possible values of X are 9 and 3.
13. 7/6, 17/36, 49/36. A tree diagram would be useful.
14. (i) Not independent; $P(Y = 1 \text{ and } X = -1) \neq P(Y = 1).P(X = -1)$.
    (iii) $p = 1/6$, $q = 1/12$.

X	-1	0	1
$P(X/Y = 1)$	0	1	0
$P(X/Y = 2)$	2/5	1/5	2/5
$P(X/Y = 0)$	2/5	1/5	2/5

15. (i) 3.5; (ii) 37.9. $\text{Var}(W) = 13 . \text{Var}(X)$. (iii) 0; (iv) 2.92.
16. (b) (i) 1/14; (ii) £4.80; (c) 25, £5.00. Differentiate $20(2n + 1) - 20 - 0.8n^2$.

Chapter 5

1. 0.4, 0.616, 113.
2. (i) 7.5; (ii) 2.37.
3. 1, 0.894; (i) 5; (ii) 0.2.
4. (b) 0.00123.
5. (i) 0.118; (ii) 0.198; 0.00402.
6. (a) 8.4, 2.4 (b) 0.5316.
7. (i) 0.0015 (ii) 0.066 (iii) 6 (iv) £15 600 (v) 11.
8. 3, 0.223, 0.988.
9. (a) (i) 0.135 (ii) 0.947 (b) (i) 0.144 (ii) 0.111.
10. $\lambda_1 + \lambda_2$ (i) 0.014 (ii) 0.039 (iii) 0.030.
11. (b) (i) 0.993; (ii) 0.944; (iii) 7, 7; (iv) 16.2, 43.8.
13. (i) 0.987; (ii) 0.514, 0.187.
14. (i) 0.325; (ii) 0.221, 3.5, 0.321.

Chapter 6

1. 2/9, 2.5, 0.45, 20/27.
2. 9/7.
3. 0.714.
4. 50.9 secs.
5. 1/4, 23/12, 2.

$$F(x) = \begin{cases} 0, & x < 0 \\ x^2/8 & 0 \leqslant x \leqslant 2 \\ (x-1)/2 & 2 < x \leqslant 3 \\ 1 & x > 3 \end{cases}$$

6. (i) 3/32; (ii) 11/16; (iii) $f(x) = \begin{cases} 3(4-x^2)/32 & -2 \leqslant x \leqslant 2 \\ 0 & \text{otherwise} \end{cases}$

   (iv) 0, 4/5.
7. (i) 8/3; (ii) $\sqrt{14/9}$; (iii) 2.54.
8. (i) 2/5; (ii) 19/15, 5/4; (iv) 0.207.
9. (ii) 1/2, 1/4. See MGF, Chapter 8.
10. $A = k$. (i) 0.232; (ii) 0.103. $P(X < 1) \cap P(X > 6)$.
11. (i) 3/4; (iii) $F(x) = \begin{cases} 0 & x < 0 \\ \frac{3}{4}\left(x^2 - \frac{x^3}{3}\right) & 0 \leqslant x \leqslant 2 \\ 1 & x > 2 \end{cases}$

    (iv) range (0, 1), $k = 24$. Find $G(y)$ then differentiate to find $g(y)$.

Chapter 7

1. 1.2.
2. $\mu = 0.693$; $\sigma = 0.0472$; 14.53%.
3. (a) 322; (b) 27; 274.
4. 0.0886.
5. 0.281. Standardise 5.5.
6. 0.322; 0.916.

7. 0.13. Use the Normal approximation to the Poisson.
8. (a) (i) 0.242, (ii) 0.145. Variance of total mass = 12. $0.8^2 + 0.62^2$. (b) £1624.
9. (i) 0.142; (ii) 0.478; (iii) 0.648.
10. $\bar{x} = 14.99$, $s = 0.368$, 82.6%.
11. (i) 1; (iii) 0.154; (iv) 0.0104; (v) 50.
12. (i) 0.0228; (ii) 0.0216; (iii) Process A.
13. (i) 0.31; (ii) 0.037; (iii) 0.0005.
14. (i) 0.460; (ii) 248.9; (iii) 0.579; $\mu = 1000.8$, $\sigma = 4$. (iv) $\mu = 24.3$ g, $\sigma = 1.5$ g. For the s.d., use $2.5^2 = x^2 + 2^2$.
15. (a) 0.685; (b) 40.5 m; (c) 40 m, 4.8 m; (d) 0.575; (e) $P(Y > 48) = 0.00023$, $P(G > 48) = 0.0475$, so choose Gwen.

Chapter 8

1. $a = 1/4$, $b = 1/4$, $c = 1/2$.
2. Look at $t(1 - t^6)/(1 - t)$ and $t(1 - t^4)/(1 - t)$.
3. 27/7. Find the PGFs for duellist A and duellist B and add the results.
4. $1/(1 - e^{-\lambda})$. Use $k(\lambda e^{-\lambda} + \lambda^2 e^{-\lambda}/2! + \ldots) = 1 \Rightarrow k(1 - e^{-\lambda}) = 1$.
5. (i) 0; (ii) n. Use $X = 2H - n$ where H gives number of heads in PGF of binomial distribution, $13n/4$ use $Y = 3H' - n$, then $G_{X+Y}(t) = G_X(t)G_Y(t)$.
6. $q^{n-1}p$, $1/p$, 36.
7. $(e^{2t} - e^{-2t})/4t$.
9. $a + 1/\lambda$, $1/\lambda^2$.
10. (i) α^k (ii) $\dfrac{(1 - \alpha)e^t}{1 - \alpha e^t}$

    $(1 - \beta\gamma)(\beta\gamma)^{k-1}$, $\dfrac{(1 - \beta\gamma)e^t}{1 - \beta\gamma e^t}$
11. 3, $e^{\sigma^2 t^2/2}$, $k = 1/2\sigma$. Use property (ii) of MGFs.

Chapter 9

1. (i) 12.96 ± 0.43; (ii) 12.96 ± 0.57.
2. 0.24 ± 0.078.
3. 78.
4. 0.0455.
5. 0.07 ± 0.04.
6. $2/\sqrt{n}$, 2171.
7. $\bar{x} = 78.4$, $s^2 = 40.64$, 78.4 ± 1.96, 0.236. Standard deviation of total mass $= \sqrt{(8 \times 40^2)}$
8. 0.105 ± 0.025. Population not infinite and sampling is without replacement. $769 \rightarrow 1250$.
9. (0.881, 0.919), 24 000, $2.58\sqrt{(0.9 \,.\, 0.1)/n} < 0.005$.
10. (0.29, 1.03).
11. (i) 0.71; (ii) 0.692.
12. (a) (i) 0.0968; (ii) 0.953; (b) (471, 559); (c) 0.209; (d) 515.6 ± 16.7; (e) 0.125 ± 0.025.
13. (b) 0.974; (c) 12; (d) $\bar{x} = \frac{3}{4}\bar{x}_A + \frac{1}{4}\bar{x}_B$; 0.585.
14. (iv) 0.0178.

Chapter 10

1. $z = -1.48$ not significant.
2. $z = 1.69$ one-tail, just significant.

3. $z = 0.62$ not significant. Pool the sample proportions.
4. $z = -2.72$, highly significant.
5. Type I error = 1/15. Type II error = 2/3.
6. $\bar{x} = 48.4, s = 17.6, z = -1.28$ not significant.
7. 0.0036
   (5.0009, 5.0151) significant difference.
8. $\bar{x} = 11.642, s = 1.892, z = -1.88$, not significant; $m = 12 + 4.66/\sqrt{n}$.
9. (a) $z = -2.77$ highly significant ($< 1\%$); (b) $z = -3.09$ highly significant ($<1\%$).
10. $0.30 < p < 0.38$. $p = 0.309, z = -2.02$. Pool the sample proportions.
11. 0.36; 0.02; $z = 1.74$, significant, one-tailed test.
12. Fair die: $\mu = 3.5, \sigma^2 = 2.92$. Biased die: $\mu = 3.9, \sigma^2 = 2.49$, 0.123.
13. (a) 0.0139; (b) 0.337; (c) $X < 6.6$; (d) $X < 10.4$; (e) 0.046; (f) 0.72.
14. 1/1200, $z = -0.95$ not significant. For the first part, use MGFs. Last part, one-tailed test.
15. $z = 2.23$ significant at the 5% level. Distribution of functions a uniform variate (see chapter 6).

Chapter 11

1. (a) 0.866; (b) (i) 2 mg^2; (ii) ($\pm$ 1.01) mg.
2. (a) (i) £6.07, (ii) £3.00; (b) (£3.26, £8.88); (c) accept hypothesis.
3. (i) (5.14, 10.86); (ii) (4.03, 19.97).
4. 40; not significant.
5. Support for greater mean, see Worked Example 11.5.
6. Support for greater mean, (0.006, 0.18) kg.
7. (i) Increase significant. See Worked Example 11.7, one-tailed test. (ii) Yes, two-tailed test.
8. (a) (23.59, 27.61); (b) difference significant, not the same.
9. (i) No deterioration at 5% level; (ii) (–1.08, 0.174).
10. Not significant, two-tailed test; significant, one-tailed test.

Chapter 12

Values of E have been rounded to 1 decimal place

1. $\chi^2 = 9.7$ with 2 degrees of freedom; highly significant ($< 1\%$).
2. $\chi^2 = 3.845$ with 1 degree of freedom; significant at 5%. Yate's correction factor.
4. $\chi^2 = 4.333$ with 2 degrees of freedom; not significant.
5. $\chi^2 = 6.288$ with 5 degrees of freedom; not significant $\Rightarrow$ reasonable fit. Integrate between respective limits giving E's to 1 d.p.
6. np; Mean = 1.6 $\Rightarrow p = 0.32$; 7.3, 17.1, 16.1, 7.6, 1.8, 0.2; $\chi^2 = 1.74$ with 2 degrees of freedom, $\Rightarrow$ reasonable fit. The last 3 classes should be combined.
7. (0.982, 4.565). Note 10 degrees of freedom when determining values of χ^2.
8. (a) $\chi^2 = 8.76$ with 2 degrees of freedom; significant $\Rightarrow$ dependent on shift.
   (b) $\chi^2 = 0.96$; not significant $\Rightarrow$ accept hypothesis.
9. $\chi^2 = 8.379$ with 6 degrees of freedom; not significant $\Rightarrow$ no association;
   $\chi^2 = 8.015$ with one degree of freedom; highly significant
   Yate's correction factor.
10. $n, 3n, 3n, n, \chi^2 = 5.83$ with 3 degrees of freedom. Yes.

Chapter 13

1. (c) -0.896.
2. (i) $R = 0.903$; (ii) $\tau = 0.711$.
4. (b) $r = -0.802$, conclusion valid ($t = 4.03$, highly significant ($< 1\%$)).

5. $r = 0.903$, highly significant ($t = 5.95$).
6. $r = 0.610$, $t = 2.78$, significant at 5%.
7. $R = -0.0875$, $\tau = -0.083$. Do not use A against B.
8. $R_{\text{buyer}} = 0.878$, $R_{\text{designer}} = 0.758$; $\tau_{\text{buyer}} = 0.733$; $\tau_{\text{designer}} = 0.556$; $R_{\text{compare}} = 0.394$; $\tau_{\text{compare}} = 0.289$. Combining the buyer's and designer's forecasts.
9. For the first part you will find it useful to use $\sum_{i=1}^{n} x_i = n(n+1)/2$ and $\sum_{i=1}^{n} x_i^2 = n(n+1)(2n+1)/6$. $R = 0.267$, $\tau = 0.222$, no indication of correlation.
10. (b) 0.937; (c) 0.973; (d) rank correlation remains unaltered but product moment correlation increases.

Chapter 14

1. $X = 2x - 17$, $y = 4.7X + 128.5$, $y = 48.6 + 9.4x$.
2. (b) $y = 237 + 3.02x$; good value – J, poor value – A, E, K.
3. Number of years protection = $0.41 + 14.2x$, thickness of paint = 0.68 mm. A more reliable predictor would be the regression line of thickness of paint on number of years of protection which gives the estimated thickness of 0.61 mm.
4. (i) 13.79; (ii) 0.951.
5. (a) $M = 11.2 + 4.3T$; (b) $m = 89.7 + 2.15t$. 124 kg.
6. (a) (i) $\bar{x} = 5$, $\bar{y} = 3$, (ii) $r = 0.894$; (b) $x_5 = 4$, $y_5 = -1$. Use Normal equations.
7. $y = 51.2 + 0.032x$. (i) (50.8, 51.6); (ii) (–0.23, 0.87). Remember to combine 2 variances.
8. (i) $\bar{y} = 10.4$; (ii) (12.8, 17.2); (iii) (4.3, 9.5). Combine 2 variances.
9. (b) $a = 3710$, $b = 192$; (d) (i) $a = 4210$, $b = 192$, (ii) $a = 4010$, $b = 207$, (iii) $a = 4160$, $b = 199$; (e) leave out people who work away from home.

Chapter 15

1. $P(\text{-ve signs} \leqslant 3) = 0.113$; one-tailed test – result not significant.
2. $T = 15\frac{1}{2}$ for 10 pairs – one-tailed test – result not significant.
3. $R = 84.5$ or $U = 70.5$, result not significant.
4. (b) (i) H_0: alcohol has no effect on the accuracy of the throwing of a dart; H_1: alcohol has an effect on the accuracy of the throwing of a dart.
   (ii) 0.8%, 7.0%, 28.9%.
   (iii) Not significant.
5. $F^4_{17} = 4.00$, which is significant.
6. $F^4_8 = 2.08$ not significant ⇒ no significant difference between technicians. $F^2_8 = 10.1$ highly significant (< 1%) ⇒ highly significant difference between arrangements. Paired comparison test between A and C useful.
7. (a) $F^2_{18} = 4.51$ which is significant. (b) Between columns: $F^2_2 = 4.11$ which is not significant. Between rows: $F^1_2 = 36.9$ which is significant.
8. (i)

	Hotel	Gallery	Church	Tower
Hotel	0	2/3	1/3	0
Gallery	2/5	0	1/5	2/5
Church	1/3	1/3	0	1/3
Tower	0	2/3	1/3	0

(ii) 37/135. Work out the top row of P^2 and multiply by the third column of P. (iii) 5/14.

9. (i) $(x^2 - x - y + 1, 2xy, y^2, 0, x + y - 2xy - x^2 - y^2)$; (iii) $x = 1/2$, $y = 1/4$, 19/4.

10.

	A	B	C	D
A	0	4/7	3/7	0
B	2/3	0	0	1/3
C	1/2	0	0	1/2
D	0	2/5	3/5	0

R after two moves $(\frac{25}{42}, 0, 0, \frac{17}{42})$; D after two moves $(0, \frac{18}{35}, \frac{17}{35}, 0)$.

On odd moves, Robert is in either C or B whereas Daniel is in A or D. On even moves, Robert is in either A or D whereas Daniel is in C or B.

Index